MIDDLE GRADES MATHEMATICS PROJECT

Factors and Multiples

William Fitzgerald

Mary Jean Winter

Glenda Lappan

Elizabeth Phillips

Addison-Wesley Publishing Company

Menlo Park, California • Reading, Massachusetts • Don Mills, Ontario
Wokingham, England • Amsterdam • Sydney • Singapore
Tokyo • Mexico City • Bogotá • Santiago • San Juan

This book is published by the Addison-Wesley Innovative Division.

ISBN 0-201-21475-x

17 18 19 20 - ML - 99

About the authors

William Fitzgerald, Ph.D. in mathematics education, University of Michigan, joined the faculty of Michigan State University in 1966 and has been Professor of Mathematics and Education since 1971. He has had extensive experience at all levels of mathematics teaching and has been involved in the development of mathematics laboratories.

Glenda Lappan, B.A., Mercer University, Macon, Georgia, and Ed.D., University of Georgia, is Professor of Mathematics at Michigan State University. She directed the mathematics component of MSU Sloan Foundation Minority Engineering Project. She has taught high school mathematics and since 1976 has worked regularly with students and teachers of grades 3–8.

Elizabeth Phillips, B.S. in mathematics and chemistry, Wisconsin State University, and M.S. in mathematics, University of Notre Dame, was visiting scholar in mathematics education at Cambridge University, England. She conducts inservice workshops for teachers and is the author of several papers and books. Currently she is a faculty member in the Department of Mathematics at Michigan State University.

Janet Shroyer, B.S., Knox College, and Ph.D., Michigan State University, has taught mathematics in Lansing public schools and at Oregon College of Education. She was a consultant in the Office of Research Services, evaluator of a teacher corps project, and a research intern in the Institute for Research on Teaching. Presently she is Associate Professor in the Mathematics Department of Aquinas College, Grand Rapids, Michigan.

Mary Jean Winter, A.B., Vassar College, and Ph.D. in mathematics, Carnegie Institute of Technology, has been Professor of Mathematics at Michigan State University since 1965. She has been involved in mathematics education at both school and college (teacher training) level since 1975. She has been especially interested in developing middle school and secondary activities using computers and other manipulatives.

A special note of recognition

Sincere appreciation is expressed to the following persons for their significant contribution to the Middle Grades Mathematics Project.

Assistants:	**David Ben-Haim** **Alex Friedlander** **Zaccheaus Oguntebi** **Patricia Yarbrough**
Consultant for evaluation:	**Richard Shumway**
Consultants for development:	**Keith Hamann** **John Wagner**

Contents

Factors and Multiples

The Middle Grades Mathematics Project (MGMP) is a curriculum program developed at Michigan State University funded by the National Science Foundation to develop units of high quality mathematics instruction for grades 5 through 8. Each unit

- is based on a related collection of important mathematical ideas
- provides a carefully sequenced set of activities that leads to an understanding of the mathematical challenges
- helps the teacher foster a problem-solving atmosphere in the classroom
- uses concrete manipulatives where appropriate to help provide the transition from concrete to abstract thinking
- utilizes an instructional model that consists of three phases: launch, explore, and summarize
- provides a carefully developed instructional guide for the teacher
- requires two to three weeks of instructional time

The goal of the MGMP materials is to help students develop a deep, lasting understanding of the mathematical concepts and strategies studied. Rather than attempting to break the curriculum into small bits to be learned in isolation from each other, MGMP materials concentrate on a cluster of important ideas and the relationships that exist among these ideas. Where possible the ideas are embodied in concrete models to assist students in moving from the concrete stage to more abstract reasoning.

THE INSTRUCTIONAL MODEL: LAUNCH, EXPLORE, AND SUMMARIZE

Many of the activities in the MGMP are built around a specific mathematical challenge. The instructional model used in all five units focuses on helping students solve the mathematical challenge. The instruction is divided into three phases.

During the first phase the teacher *launches* the challenge. The launching consists of introducing new concepts, clarifying definitions, reviewing old concepts, and issuing the challenge.

The second phase of instruction is the class *exploration*. During exploration, students work individually or in small groups. Students may be gathering data, sharing ideas, looking for patterns, making conjectures, or developing other types of problem-solving strategies. It is inevitable that students will exhibit variation in progress. The teacher's role during exploration is to move about the classroom, observing individual performances and encouraging on-task behavior. The teacher urges students to persevere in seeking a solution to the challenge. The teacher does this by asking appropriate questions and by providing confirmation and redirection where needed. For the more able students, the teacher provides extra challenges related to the ideas being studied. The extent to which students require attention will vary, as will the nature of attention they need, but the teacher's continued presence and interest in what they are doing is critical.

When most of the students have gathered sufficient data, they return to a whole class mode (often beginning the next day) for the final phase of instruction, *summarizing*. Here the teacher has an opportunity to demonstrate ways to organize data so that patterns and related rules become more obvious. Discussing the strategies used by students helps the teacher to guide them in refining these strategies into efficient, effective problem-solving techniques.

The teacher plays a central role in this instructional model. The teacher provides and motivates the challenge and then joins the students in exploring the problem. The teacher asks appropriate questions, encouraging and redirecting where needed. Finally, through the summary, the teacher helps students to deepen their understanding of both the mathematical ideas involved in the challenge and the strategies used to solve it.

To aid the teacher in using the instructional model, a detailed instructional guide is provided for each activity. The preliminary pages contain a rationale; an overview of the main ideas; goals for the students; and a list of materials and worksheets. Then a script is provided to help the teacher teach each phase of the instructional model. Each page of the script is divided into three columns:

TEACHER ACTION	TEACHER TALK	EXPECTED RESPONSE
This column includes materials used, what to display on the overhead, when to explain a concept, when to ask a question, etc.	This column includes important questions and explanations that are needed to develop understandings and problem-solving skills, etc.	This column includes correct responses as well as frequent incorrect responses and suggestions for handling them.

Worksheet answers, when appropriate, and review problem answers are provided at the end of each unit; and for each unit test, an answer key and a blackline master answer sheet is included.

RATIONALE

Some relationships are so basic to an area of study that they earn the name *fundamental.* One such relationship is the Prime Factorization Theorem, also called the Fundamental Theorem of Arithmetic. It states that any natural number except 1 can be written as a product of primes in one and only one way, except possibly for the order of factors. The prime numbers are the building blocks for further study of the properties and the many relationships among the natural numbers. The Factors and Multiples Unit focuses on this fundamental theorem and related ideas such as factor, divisor, multiple, common factor, common multiple, relatively prime, and composite.

Introduction

These ideas are a part of the scope and sequence for any upper elementary or middle school mathematics curriculum. However, the student very often meets them piecemeal—a little here, a little there—and may never make the necessary connections among the relationships. This unit provides a deeper understanding of the concepts and processes involved in finding factors and multiples. Since little in the way of computational skill is involved, students can discover these relationships by solving a variety of problems and at the same time develop their problem-solving skills.

The study of fractions can naturally follow this unit, because a sound understanding of factors and multiples is a prerequisite for successfully understanding and operating with fractions.

Not only are these ideas fundamental to success with fractions but also to the study of algebra. Algebra is a brief and general language used to describe patterns, in particular, arithmetic patterns. The relationship of factor to multiple is basic to an understanding of the language of algebra. A simple expression like $6x$ must be seen as a product of two factors, 6 and x, and as a multiple of 6 and of x before a student can meaningfully distinguish between the expression, $6x$, and $6 + x$.

These arithmetic ideas are further extended to algebra when students factor and multiply algebraic expressions. If students acquire a language in which to talk and read about these ideas as they are developing computational skills, applying the ideas in new situations becomes easier. The Factors and Multiples Unit provides a context for understanding both the concepts and the language of factors and multiples.

UNIT OVERVIEW

Students' interest and motivation are easily captured by the games and puzzles used to teach concepts and problem-solving strategies in this unit. The eight activities fall naturally into three parts. The first three activities focus on the basic relationship between factors and products and on the process of finding all the factor pairs of a number. The factor pairs of a number are used to describe rectangles whose area is the given number and whose dimensions are the factor pair. These rectangles are put on a grid to help students discover how far to check to make sure they have all the factor pairs of a number. For any number N, it is only necessary to check all the numbers 1 through M where M is the largest number such that $M \times M \leq N$.

The second set of activities begins by asking students to find the prime factorization of a number and then to use the prime factorization to find the least common multiple and the greatest common factor of a pair of numbers.

The last two activities apply the concepts of prime, factor, common multiples, and common factors to new situations. This set starts with the "Sieve of Eratosthenes," which is a method of finding primes by casting out multiples of primes. The last activity is a game of paper pool that requires students to analyze several patterns by applying the ideas from the previous activities.

Activity 1

THE FACTOR GAME

OVERVIEW

The Factor Game is a two-person game that provides a rich set of experiences with the numbers from 1 to 30 and their *factors.* One student selects a number and the other student finds all the proper factors of the number. This process alternates between the two players until there are no factors left for the remaining numbers. Scores for each person are obtained by adding the numbers selected by each player. An analysis of good first moves leads to a definition of *prime* and *composite numbers.* Numbers that have many factors are not good first moves because the sum of their proper factors is larger than the number. In mathematics these are called *abundant numbers.* Good first moves are numbers in which the sum of the proper factors is less than the number. These numbers are called *deficient numbers.* There are two numbers between 1 and 30 that are neither good nor not good first moves. The numbers 6 and 28 are exactly equal to the sum of their proper factors and in mathematical terms are called *perfect numbers.*

The classification of numbers in the various ways described above comes naturally out of a discussion of the Factor Game. In addition, the Factor Game provides practice in basic multiplication and division facts. The relationship between *factor* and *product* becomes clear. Using the strategy of dividing to check to see whether a number is a factor of another number reinforces that *divisor* and *factor* mean the same thing.

Students enjoy the Factor Game and benefit from playing it many times. They should be encouraged to play a round at odd times during the day. You may wish to establish a tournament to extend a longer period of time. The class might be divided into teams and records of scores and standings kept. Teams might hold strategy sessions where team members help each other improve their factor skills.

Goals for students

1. Learn the definition of *factor* and *divisor.*
2. Quickly find the factors of whole numbers 1 to 30.
3. Learn the definition of *prime* and *composite, abundant* and *deficient numbers.*
4. Classify whole numbers greater than one as prime or composite.
5. Develop best move strategies for the Factor Game by looking at abundant and deficient numbers.

Materials

Calculators.
Colored pens or crayons.
Colored transparency pens (2).
*Factor Game (Materials 1-1).

Worksheets

1-1, 30-Game Board
*1-2, Analyzing First Moves for the 30-Game Board.
1-3, Moves for the 49-Game Board
1-4, Practice Problems.

Transparencies

Starred items should be made into transparencies.

THE FACTOR GAME

TEACHER ACTION	TEACHER TALK	EXPECTED RESPONSE
Review the concept of a factor.	What is a factor of a number?	When you multiply it by something, you get the number.
Ask for each factor.	What are the factors of 12?	Elicit 1, 2, 3, 4, 6, 12.
	Tell me how you know that each of these is a factor.	$1 \times 12 = 12$; $2 \times 6 = 12$; $3 \times 4 = 12$; $4 \times 3 = 12$; $6 \times 2 = 12$; $12 \times 1 = 12$.
	What happens if you divide 12 by one of its factors? Use your calculators.	It comes out even; you get another factor.
	What happens if you divide 12 by a number that's not a factor? Try $12 \div 7$.	Various answers. Elicit that there is something after the decimal point.
	Another name for factor is *divisor.*	
	Is 13 a divisor of 29?	No.
	Is 13 a divisor of 39?	Yes.
	What are the divisors of 18?	1, 2, 3, 6, 9, 18.
Begin the game. Display a transparency of the Factor Game (Materials 1-1) on the overhead.	We are going to learn to play a new game called the Factor Game. Two people play, taking turns. Player A selects a number, then player B selects all the divisors of player A's number.	
	Then player B selects a new number and player A selects all the factors of player B's number.	
	There are some other rules that I'll explain as we play.	
	We'll play the first game as a group—class vs. teacher.	
Select a number.	I'll go first.	
Circle 26 with green.	I picked 26. You now circle all the factors of 26 that you can find that haven't already been circled.	

TEACHER ACTION	TEACHER TALK	EXPECTED RESPONSE
Circle numbers named by the class in red.	What numbers do you get?	1, 2, 13.
	The score we get is the sum of the numbers we have circled. I have 26 points.	
	You have 1 + 2 + 13 = 16 points so far.	
Call on the class for a move.	Now you select a number, and I circle the factors.	
Continue playing until the class selects a number that *has no uncircled factors left.*	Here's another rule. You selected a number with no factors left. That's an illegal move. (A move is legal only if there's a factor for the other player to get.)	
Example: The board is in this state and the class selects 16.	*You may make an illegal move* and you get to add the number in your score, *but* you lose your next turn to select a number.	
	If you make an illegal move and then right away I make an illegal move, we just continue.	
Continue until end of game.	The game is over when there are no more legal moves left on the board.	
	The winner is the player with the larger score.	
Add up the class score and your score to illustrate the scoring scheme.		
Use colored pens or different schemes to mark numbers so that at the end of the game each player knows which numbers are his or hers.		

①	②	③	④	⑤
6	7	⑧	9	10
11	12	13	14	⑮
16	⑰	18	19	20

TEACHER ACTION	TEACHER TALK	EXPECTED RESPONSE
Students need calculators; Worksheet 1-1, 30-Game Board; and colored pens (a different color for each student in a pair).	Play the game in pairs. Take turns starting first. Do the games on the 30-Game boards first.	
If students ask what to do if a player misses a factor, tell them to leave it uncircled for future plays.		
Allow time for all students to play at least three times before you summarize.		
The 49-Game Board can be used as an extra challenge at any stage.		

TEACHER ACTION	TEACHER TALK	EXPECTED RESPONSE
Students need Worksheet 1-2, Analyzing First Moves for the 30-Game Board.	Suppose you are starting a round of the Factor Game and you make the first move.	
	If you take 26, how many points does the other player get?	$1 + 2 + 13 = 16$
	What about 18 as a first move? What does the opponent get?	$1 + 2 + 3 + 6 + 9 = 21$
	Is choosing 18 a good first move? Why?	No; your opponent scores more than you do.
Put a transparency of Worksheet 1-2, Analyzing First Moves for the 30-Game Board, on the overhead.	Let's look at each possible first move, calculate what the opponent scores, and record it like this: 26–16 means that when 26 is the first move, the other player gets 16.	
Illustrate.	On your Analyzing First Moves Worksheet, record the results we have so far for 26 and 18.	

Factors			First Moves	
Number Picked	Opponent Gets	Sum	Good	Not Good
⋮				
18	1,2,3,6,9	21		18–21
⋮				
26	1,2,13	16	26–16	

TEACHER ACTION	TEACHER TALK	EXPECTED RESPONSE
As an alternative, break the class into groups and assign each group part of the numbers to analyze.	Now complete the table.	
Give students time to complete the analysis. Then make a complete record on the overhead so that every student will have a correct table of moves.		

TEACHER ACTION	TEACHER TALK	EXPECTED RESPONSE
With 6 and 28, suggest that they be listed between the Good–Not Good columns as shown below. 6 | 6 28 | 28		
Ask.	Now let's analyze our table. What first moves give the other player exactly 1 point?	2, 3, 5, 7, 11, 13, 17, 19, 23, 29.
	What are these numbers called?	Prime numbers.
	A *prime* is a number that has exactly two factors, itself and 1.	
	All other numbers except the number 1 are called *composite* numbers.	
	Name some composite numbers. What is the fewest number of factors that a composite number can have?	Various answers. Three; examples of composite numbers with three factors are 4, 9, 25.
	How many factors does the number 1 have?	One.
	1 is neither a prime number nor a composite number. It is the only number with exactly one factor.	
Ask.	Look at your table again. Were all the prime numbers good first moves?	Yes.
	Were all the composites good first moves?	No; some were good, some were not good.
	What was the best first move?	29.
	What was the worst first move?	24 or 30.

TEACHER ACTION	TEACHER TALK	EXPECTED RESPONSE
Explain.	The ancient Greek mathematicians had another way to classify numbers. They called one group *abundant*, one group *deficient*, and one group *perfect* by looking at the proper factors of a number. The *proper factors* are all the factors that are less than the number.	
	If the sum of the proper factors is larger than the number, the number is called an *abundant* number. Which of our group are the abundant numbers?	The not good first moves.
	The *perfect* numbers exactly equal the sum of their proper factors. That means 6 and 28 are perfect.	
As an extra challenge, have students find the next perfect number. (496 is the next one!)	This leaves the word *deficient* to describe the good moves.	
	Why does this word fit?	The sum of the proper factors is smaller than the number; these numbers have few factors.
List the sums students provide for each first move.	Look at these numbers for first moves: 2, 4, 8, 16, 32.	
Number — Sums 2 – 1 4 – 3 8 – 7 16 – 15 32 – 31	In each case, what does the other player get?	2–1, 4–3, 8–7, 16–15, 32–31.
	If we played on a 100-board, how many points would your opponent get if you took 64?	64–63.
	These are almost perfect numbers or barely smart first moves.	
	What is the next almost perfect number?	128.

TEACHER ACTION	TEACHER TALK	EXPECTED RESPONSE
Explain.	Consider this situation: George and Sally are playing the Factor Game. The teacher walks by, glances at the board and sees that the numbers from 1–15 are all circled. The teacher immediately says "There are no legal moves left." Is the teacher correct?	Yes. If the numbers from 1–15 are circled already, there are no factors left for any of the numbers from 16–30.
	We can use this as a quick check to see if a game is over. Some games will end *before* the numbers from 1 to 15 are all circled.	If the students do not see this, ask for the largest factor the opponent can get when you circle 30. 29? 28? etc.
Ask.	Did anyone have all the numbers circled when the game ended?	
	If you could do it, what would be the total score for both players?	465.
	Play the 30-Game two more times. Play it once to see whether you can get all the numbers circled before the game ends.	
Post the fewest moves on the board.	Play it a second time using the fewest number of moves possible to complete a game.	
Practice problems are provided in Worksheets 1-3 and 1-4.		
Worksheet 1-3 extends the analysis of good and not good moves to a 49-Game Board.		
Worksheet 1-4 is a collection of problems emphasizing the classifications of numbers introduced in the summary.		
These practice problems could be used as a homework assignment.		

Factor Game

1	2	3	4	5
6	7	8	9	10
11	12	13	14	15
16	17	18	19	20
21	22	23	24	25
26	27	28	29	30

NAME ____________________

30–Game Board

1	2	3	4	5
6	7	8	9	10
11	12	13	14	15
16	17	18	19	20
21	22	23	24	25
26	27	28	29	30

30–Game Board

1	2	3	4	5
6	7	8	9	10
11	12	13	14	15
16	17	18	19	20
21	22	23	24	25
26	27	28	29	30

30–Game Board

1	2	3	4	5
6	7	8	9	10
11	12	13	14	15
16	17	18	19	20
21	22	23	24	25
26	27	28	29	30

NAME ______________________

30–Game Board

1	2	3	4	5
6	7	8	9	10
11	12	13	14	15
16	17	18	19	20
21	22	23	24	25
26	27	28	29	30

49–Game Board

1	2	3	4	5	6	7
8	9	10	11	12	13	14
15	16	17	18	19	20	21
22	23	24	25	26	27	28
29	30	31	32	33	34	35
36	37	38	39	40	41	42
43	44	45	46	47	48	49

NAME ____________________

Analyzing First Moves for the 30–Game Board

Factors **First Moves**

1st number picked is	Opponent Gets	SUM	GOOD	NOT GOOD
2				
3				
4				
5				
6				
7				
8				
9				
10				
11				
12				
13				
14				
15				

NAME

Analyzing First Moves for the 30–Game Board

	Factors		First Moves	
1st number picked is	Opponent Gets	SUM	GOOD	NOT GOOD
16				
17				
18				
19				
20				
21				
22				
23				
24				
25				
26				
27				
28				
29				
30				

NAME ______________________

Moves for the 49–Game Board

Play the Factor Game on the 49–Game Board. Analyze the first moves by filling in the chart below. Since the numbers 1–30 have been done for the 30–Game Board you need only to look at the numbers 31–49.

	Factors		**First Moves**	
1st number picked is	Opponent Gets	SUM	GOOD	NOT GOOD
31				
32				
33				
34				
35				
36				
37				
38				
39				
40				
41				
42				
43				
44				
45				
46				
47				
48				
49				

NAME ____________________

Practice Problems

Use the table of first moves for numbers 1 through 49 (Worksheet 1-3) to complete the following.

1. List all the prime numbers from 1 to 49.

2. List all the numbers from 1 to 49 that are abundant numbers.

3. List all the numbers from 1 to 49 that are deficient numbers.

4. List all the numbers from 1 to 49 that are perfect numbers.

5. List all the numbers that have 2 as a factor.

What do we call these numbers? ____________________

What do we call numbers that do not have 2 as a factor? ____________________

NAME ______________________

Practice Problems

6. What number(s) have the most factors?

7. What factor is paired with

16 to give 48? ______________ 16 to give 32? ______________

8. What factor is paired with

12 to give 48? ______________ 12 to give 36? ______________

9. What factor is paired with

6 to give 48? ______________ 6 to give 42? ______________

10. What factor is paired with

4 to give 48? ______________ 4 to give 44? ______________

11. What factor is paired with

2 to give 48? ______________ 2 to give 34? ______________

12. What is the *best* first move on a 49–Game Board?

13. What is the *worst* first move on a 49–Game Board?

Activity 2 CREATE A GAME

OVERVIEW

In the Factor Game the students selected a number first and then found the factors of that number. In this activity, the Product Game reverses the focus: students first find the factors and then find the product. Together the strategies for playing the two games reinforce the relationships found in the simple statement A × B = C. Given any two of the values A, B, or C, the students learn to find the missing part. If the students are given only one of the values A, B, or C, they learn to find pairs of values for the other two parts.

The Product Game is played on a 6 × 6 product matrix in which the winner must get four in a row. Once the students learn the game and play several rounds, time should be spent summarizing rules, strategies, and other general characteristics.

The challenge to create a new product game provides students with the need to think about the fundamental ideas involved. Students can be challenged to create their own games in groups or individually, beginning with simple sets of factors and advancing to more complicated games. To create an interesting game requires a careful analysis of the existing 6 × 6 game followed by an application of the analysis to a new situation. This provides excellent practice with significant problem-solving techniques.

Goals for students

1. Learn that multiplying *factors* gives a *product.*
2. Given one factor of a number, find another factor by dividing the known factor into the number.
3. Test to see if a number is a factor of a number by dividing.
4. Given a list of factors, find all possible products that can be made by selecting the factors two at a time.

Materials

Colored pens.
Markers (chips, paper clips, etc.).
List of Possible Products Beginning at 1 (Materials 2-1).

Worksheets

*2-1, 6 × 6 Product Game.
*2-2, Create a 3 × 3 Product Game.
2-3, Create a 4 × 4 Product Game.
2-4, Create a 5 × 5 Product Game.
2-5, Find the Factors.

Transparencies

Starred items should be made into transparencies.

CREATE A GAME

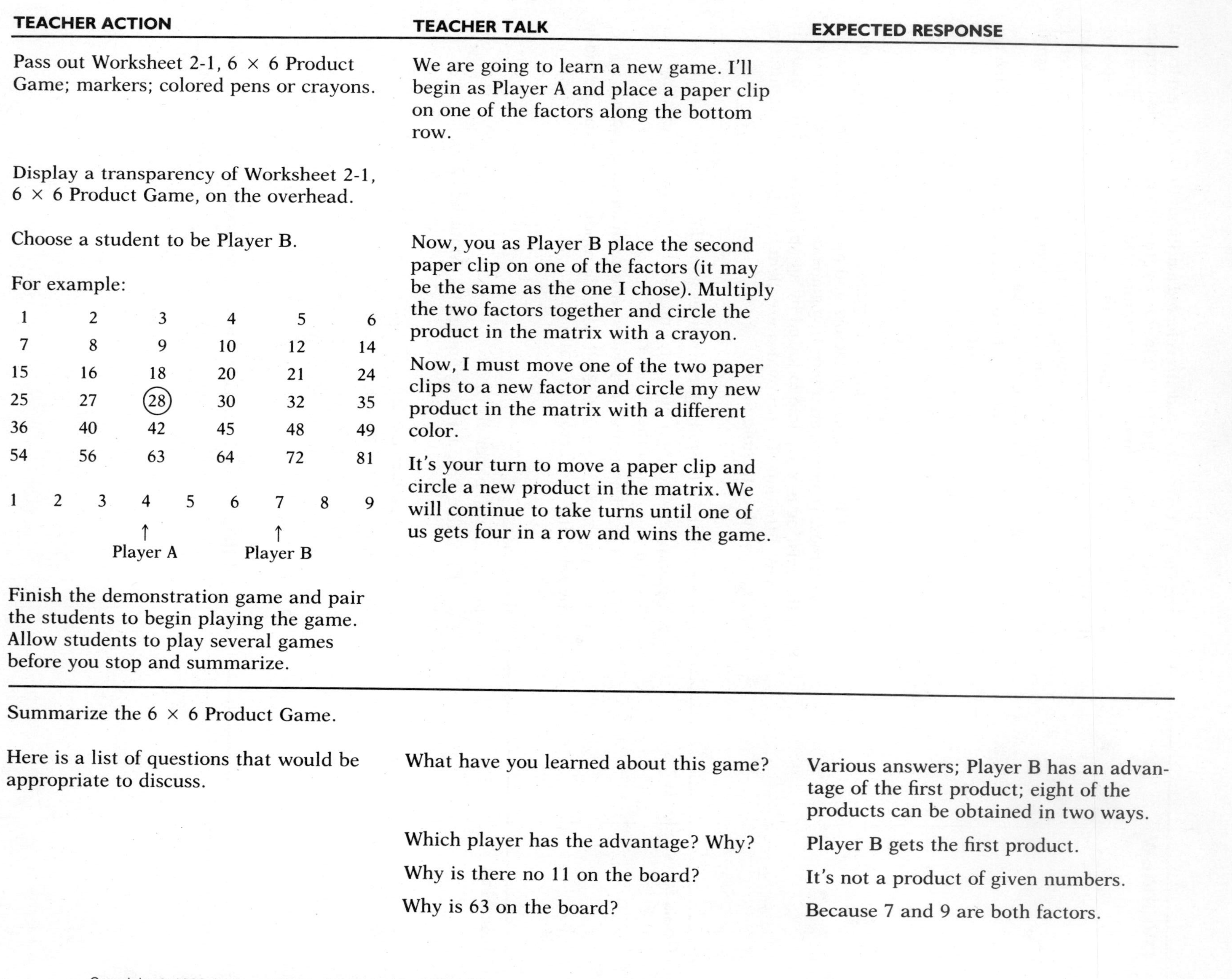

TEACHER ACTION	TEACHER TALK	EXPECTED RESPONSE
Pass out Worksheet 2-1, 6 × 6 Product Game; markers; colored pens or crayons.	We are going to learn a new game. I'll begin as Player A and place a paper clip on one of the factors along the bottom row.	
Display a transparency of Worksheet 2-1, 6 × 6 Product Game, on the overhead.		
Choose a student to be Player B.	Now, you as Player B place the second paper clip on one of the factors (it may be the same as the one I chose). Multiply the two factors together and circle the product in the matrix with a crayon.	

For example:

1	2	3	4	5	6
7	8	9	10	12	14
15	16	18	20	21	24
25	27	(28)	30	32	35
36	40	42	45	48	49
54	56	63	64	72	81

1 2 3 4 5 6 7 8 9

↑ Player A (at 4) ↑ Player B (at 7)

Now, I must move one of the two paper clips to a new factor and circle my new product in the matrix with a different color.

It's your turn to move a paper clip and circle a new product in the matrix. We will continue to take turns until one of us gets four in a row and wins the game.

Finish the demonstration game and pair the students to begin playing the game. Allow students to play several games before you stop and summarize.

TEACHER ACTION	TEACHER TALK	EXPECTED RESPONSE
Summarize the 6 × 6 Product Game.		
Here is a list of questions that would be appropriate to discuss.	What have you learned about this game?	Various answers; Player B has an advantage of the first product; eight of the products can be obtained in two ways.
	Which player has the advantage? Why?	Player B gets the first product.
	Why is there no 11 on the board?	It's not a product of given numbers.
	Why is 63 on the board?	Because 7 and 9 are both factors.

TEACHER ACTION	TEACHER TALK	EXPECTED RESPONSE
	Which products can be obtained in more than one way?	4, 6, 8, 9, 12, 16, 18, 24.
	Is every product possible on the board?	Yes.
	Why are there not 9×9 or 81 products?	For every $a \times b$ there is a $b \times a$, which reduces the number of products by half; some products can be obtained by more than one pair of factors.
	Turn your boards over so that you cannot see them. Describe the board.	Use factors from 1 to 9. The products are put into a square grid.
	What numbers are in the first row? How were they obtained?	1×1 1×2 1×3 1×4, 2×2 1×5 1×6, 2×3
	What numbers are in the second row? How were they obtained?	1×7
	What numbers are in the third row? How were they obtained?	Students continue row by row, listing all of the factors for each product on the grid.
	What numbers are in the fourth row? How were they obtained?	
	What numbers are in the fifth row? How were they obtained?	
	What numbers are in the sixth row? How were they obtained?	
Pass out Worksheet 2-2, Create a 3×3 Product Game.		
Explain.	We are going to create a new 3×3 board on which to play a product game. We need to select the factors we will use and place their products in the grid.	
	If we use all the factors from 1 to 9 how many different products can we make?	36; we used that many in the 6×6 Product Game.
	We have only nine spaces. What could happen in the play of our game if we use too many factors?	Players may not be able to circle a number on the board at their turn.

TEACHER ACTION	TEACHER TALK	EXPECTED RESPONSE
	Let's try to use only enough factors to exactly fill the board. If we use just the factor 1, what products can we make?	1.
	If we use 1 and 2, what products can we make?	1, 2, 4.
	If we use 1, 2, and 3, what products can we make?	1, 2, 3, 4, 6, 9.
Demonstrate:	Let's keep a record as we go on the 3 × 3 Product Game board. We will add only *new* products as we add more factors.	
	Now if we use 4 as a factor, what new products do we make?	8, 12, 16.
	Do we have what we need for the 3 × 3 Product Game board?	Yes; exactly nine products.
	Place the nine factors on your board.	
	How many in a row should the winner get?	Students may say two or three; three makes a nice game.
Ask.	Could we make a 2 × 2 Game Board? Why or why not?	Students may say yes or no; when they see they cannot get exactly four products, suggest blacking out one of the spaces and filling in the others with 1, 2, 4. 
	Would it make an interesting game?	No!

Factors	Possible Products
1	1
2	2, 4
3	3, 6, 9
4	8, 12, 16

TEACHER ACTION	TEACHER TALK	EXPECTED RESPONSE
Pass out Worksheet 2-3, Create a 4 × 4 Product Game.	Play a round or two of the 3 × 3 Product Game with your partner. Then working together, create a 4 × 4 Product Game on the board provided and play your game.	
Worksheet 2-4, Create a 5 × 5 Product Game, can be assigned as homework or can be used as an extra challenge.		
A list of possible products of consecutive factors for 1 to 16 is given for teacher reference (Materials 2-1). Some teachers may have the class make out their own chart to assist in making product games.		
As an extra challenge, have students create a 10 × 10 Product Game Board. On a board this big, the game becomes uninteresting. A better game is to play for the most markers on the board. (You can use a fixed time of play to end the game). The following variation can be added:		
If you can surround an opponent's marker, you can replace his or her piece with yours. The player with the most markers wins.		

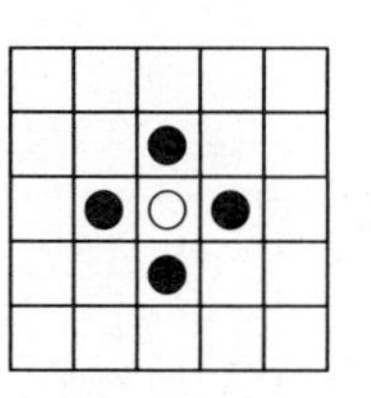

A 10 × 10 Product Game Board is provided for teacher reference (Materials 2-2).

TEACHER ACTION	TEACHER TALK	EXPECTED RESPONSE
Allow students to share any interesting games they have created.		
Pass out Worksheet 2-5, Find the Factors. This sheet can also be given as homework.		
Explain.	We do not necessarily have to use consecutive numbers beginning with 1 as the factors of our games. We do want the game board to contain *every* possible product of the factors we choose.	
Allow students time to work and then check answers.	Worksheet 2-5 has two game boards that do not have consecutive factors. Find the factors used to make the board, and find the product that is missing.	

List of Possible Products Beginning at 1

Factors Starting With 1	Possible Products	Number of Products Added	Total Number of Products
1	1	1	1
2	2, 4	2	3
3	3, 6, 9	3	6
4	8, 12, 16	3	9
5	5, 10, 15, 20, 25	5	14
6	18, 24, 30, 36	4	18
7	7, 14, 21, 28, 35, 42, 49	7	25
8	32, 40, 48, 56, 64	5	30
9	27, 45, 54, 63, 72, 81	6	36
10	50, 60, 70, 80, 90, 100	6	42
11	11, 22, 33, 44, 55, 66, 77, 88, 99, 110, 121	11	53
12	84, 96, 108, 120, 132, 144	6	59
13	13, 26, 39, 52, 65, 78, 91, 104, 117, 130, 143, 156, 169	13	72
14	98, 112, 126, 140, 154, 168, 182, 196	8	80
15	75, 105, 135, 150, 165, 180, 195, 210, 225	9	89
16	128, 160, 176, 192, 208, 224, 240, 256	8	97

This list is provided as an aid to the teacher. By examining the Total Number of Products column, reasonable board sizes can be determined. For example, a 7 × 7 board would produce 49 spaces. The table shows that there are 42 products with the factors 1 to 10, so seven spaces should be blacked out.

To read the chart, select a row and use all the factors and products from 1 through that row. For example, the row starting with 4 gives the factors 1, 2, 3, 4, and all their products: 1; 2, 4; 3, 6, 9; and 8, 12, 16.

10 × 10 Product Game

	1	2	3	4	5	6	7	8	
9	10	11	12	13	14	15	16	18	20
21	22	24	25	26	27	28	30	32	33
35	36	39	40	42	44	45	48	49	50
52	54	55	56	60	63	64	65	66	70
72	75	77	78	80	81	84	88	90	91
96	98	99	100	104	105	108	110	112	117
120	121	126	128	130	132	135	140	143	144
150	154	156	160	165	168	169	176	180	182
192	195	196	208	210	224	225	240	256	

Factors:

1 2 3 4 5 6 7 8

9 10 11 12 13 14 15 16

NAME ______________________

6 × 6 Product Game

1	2	3	4	5	6
7	8	9	10	12	14
15	16	18	20	21	24
25	27	28	30	32	35
36	40	42	45	48	49
54	56	63	64	72	81

Winner ______________________

1	2	3	4	5	6
7	8	9	10	12	14
15	16	18	20	21	24
25	27	28	30	32	35
36	40	42	45	48	49
54	56	63	64	72	81

Winner ______________________

1	2	3	4	5	6
7	8	9	10	12	14
15	16	18	20	21	24
25	27	28	30	32	35
36	40	42	45	48	49
54	56	63	64	72	81

Winner ______________________

1	2	3	4	5	6
7	8	9	10	12	14
15	16	18	20	21	24
25	27	28	30	32	35
36	40	42	45	48	49
54	56	63	64	72	81

Winner ______________________

1 2 3 4 5 6 7 8 9

NAME ______________________

Create a 3 × 3 Product Game

Factors	Possible Products

NAME ______________________

Create a 4 × 4 Product Game

Why are the corners blacked out?

NAME

Create a 5 × 5 Product Game

NAME ____________________

Find the Factors

Find the factors used to make these Product Game boards. List the factors in the spaces provided. Find the secret number (?).

1.

	4	6
9	10	14
15	?	25
35	49	

? = ☐

_____ _____ _____ _____

2.

4	6	8	9
12	16	18	24
27	32	36	48
54	64	72	?

? = ☐

_____ _____ _____ _____ _____ _____

Activity 3 FACTOR PAIRS

OVERVIEW

In finding all the factors of 98, is it necessary to test all the numbers from 1 to 98? Is it even necessary to test all the numbers from 1 to 49 (49 = 98 ÷ 2)? The purpose of this activity is two-fold: to reiterate that factors come in pairs, and to identify the largest number that needs to be tested if all the factors are being sought. In this case the largest number that needs to be tested is 9, the whole number that is closest to the square root of 98.

In the Launch, after listing all the factor pairs for several different numbers, the students observe a cutoff or gap beyond which factor pairs appear in reverse order with reversed components. Next students identify factor pairs with rectangles of fixed area. After these rectangles are graphed, students use the symmetry of the diagonal line drawn from the origin to the outer corners of a square to predict where the cutoff will occur. The diagonal line is referred to as the *line of squares.*

Goals for students

1. Recognize that a factor pair for a number corresponds to a rectangle whose area is that number.
2. Observe that the graphs of the rectangles for a fixed number are symmetrical at the *line of squares.*
3. Identify where the graph of the factor pairs for a given number meets the line of squares.
4. Identify the largest possible factor (before the cutoff) of a given number; find all factors of that number.

Materials

Square bathroom tiles (20) (optional).

Worksheets

*3-1, Factor Pairs and Rectangles.
3-2, Crossing the Line.
3-3, Practice Problems.

Transparencies

Starred item should be made into transparencies.

FACTOR PAIRS

TEACHER ACTION	TEACHER TALK	EXPECTED RESPONSE
Pass out Worksheet 3-1, Factor Pairs and Rectangles.	$7 \times 8 = 56$. We say 7×8 is a *factor pair* for 56. Today we will concentrate on finding *all* the factor pairs for a number.	
	What are the factor pairs for 12? Tell them to me in order.	
Record a list for 12, and then 28, on the blackboard. 1×12 1×28 2×6 2×14 ⋮ ⋮		1×12 2×6 3×4 4×3 6×2 12×1
	What are the factor pairs, in order, for 28?	1×28 2×14 4×7 7×4 14×2 28×1
In the list of factor pairs draw a line between $\frac{4 \times 7}{7 \times 4}$.	If we regard 2×14 as the same pair as 14×2, where do we stop obtaining new factor pairs for 28?	*After* 4×7, or between 4×7 and 7×4.
and draw a line between $\frac{3 \times 4}{4 \times 3}$.	Where do we stop obtaining new factor pairs for 12?	*After* 3×4, or between 3×4 and 4×3.
	We call this line where the pairs reverse the cutoff.	
	List the factor pairs for 16. Where does the cutoff occur?	1×16 2×8 (4×4) (circled) 8×2 16×1 *At* 4×4.

TEACHER ACTION	TEACHER TALK	EXPECTED RESPONSE
	Sometimes the cutoff occurs *at* a factor pair; sometimes it occurs *between* two factor pairs. We are going to investigate the question "How can we locate the cutoff without listing all the factor pairs?"	If students observe that you hit the cut-off when you get a repeat, look at the pairs for 46. 1×46 2×23 23×2 46×1 We check a lot of unnecessary numbers, all those from 3 to 22, before we see the repeat at 23×2. We need a better scheme.
Record factors of 12 on a transparency of Worksheet 3-1 and draw rectangles one at a time, giving students time to draw each as you go. Pairs 1×12 2×6 ⋮ Side Edge: 6, 5, 4, 3, 2, 1 Bottom Edge: 0 1 2 3 4 5 6	To find an answer to how to locate the cutoff without listing all the factor pairs, we will think of a factor pair as representing a rectangle. 3×4 will be a rectangle with bottom edge 3 and side edge 4. Draw it on graph number 1 on your sheet like this. Now we will draw the rectangle for 2×6 on the same grid, overlapping the 3×4 rectangle. The lower left corner of *each* rectangle will be the point (0,0) on the grid.	For 5th and 6th graders the possible rectangles for a number can first be formed from square bathroom tiles and then copied onto the grid paper. Give each student or pair of students 20 square tiles. First count out 12 of them. Form a rectangle using all 12 tiles. Make a grid paper copy. Reuse the 12 tiles to form another rectangle. Copy it. Continue until all possible rectangles are found. Repeat for all 20 tiles. Students can use the tile rectangles to draw in the rectangles on the grid. You may need to review how to locate and label points on a grid.

TEACHER ACTION	TEACHER TALK	EXPECTED RESPONSE
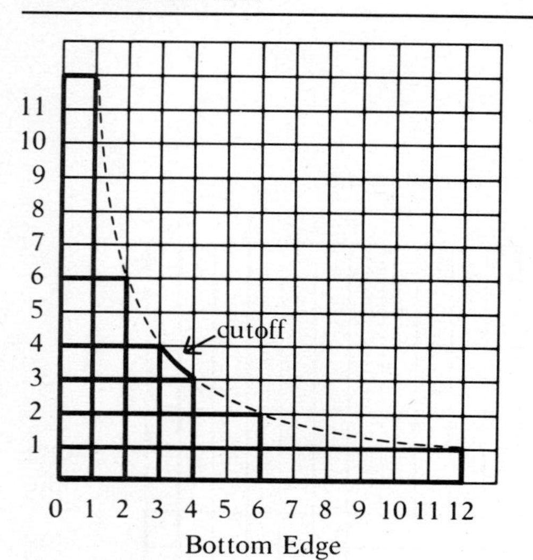	Draw the remaining rectangles.	
	Where does the cutoff occur?	Various answers.
	Where do the factor pairs reverse or turn around?	Between rectangles 3 × 4 and 4 × 3.
	If we were to draw all the rectangles for 24, how tall would the first be?	24 tall.
	That's really too long to fit on your graphs. In problems 1 to 5, list all the factor pairs for the boxed number. Draw in all the rectangles for those that fit the graph.	
Let students complete Worksheet 3-1.	Problem 6 is a little different. We are looking at *squares*. They do not have the same area.	

TEACHER ACTION

Draw squares on the grid transparency.

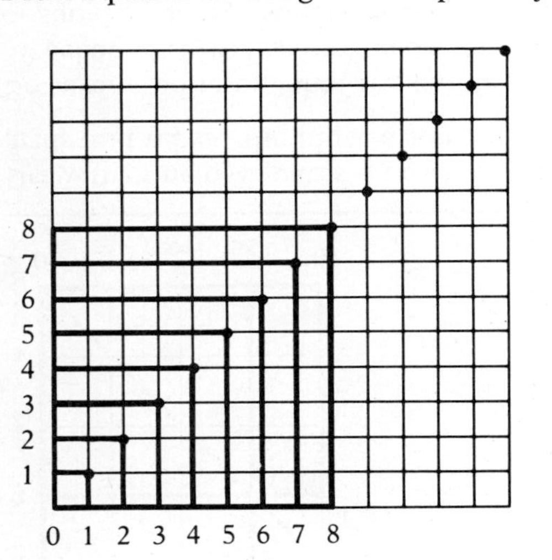

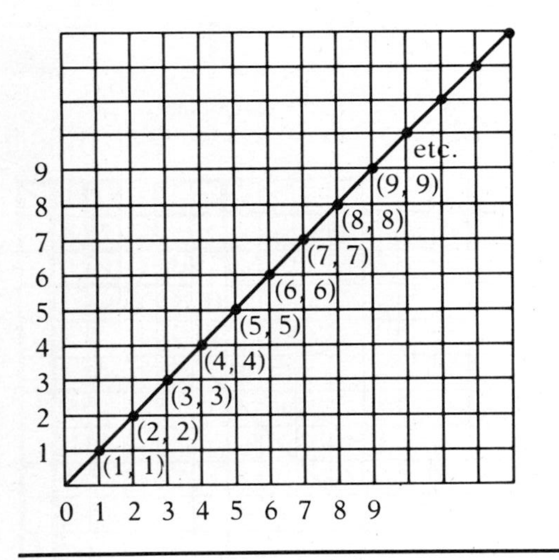

TEACHER TALK

Let's look at problem 6.

Squares are those numbers that have a factor pair with equal factors, for example:

1×1 is an equal-factor pair for 1
2×2 is an equal-factor pair for 4
3×3 is an equal-factor pair for 9
4×4 is an equal-factor pair for 16
5×5 is an equal-factor pair for 25
6×6 is an equal-factor pair for 36
7×7 is an equal-factor pair for 49
and so on.

So the numbers 1, 4, 9, 16, 25, 36, 49, and so on are called *squares* or *square numbers;* one of the factor pairs has equal factors. When we think of this factor pair as a rectangle, the bottom edge and the side edge are the same, so we have a square. Name the next three square numbers *and* their equal factor pairs.

$8 \times 8 \longrightarrow 64$
$9 \times 9 \longrightarrow 81$
$10 \times 10 \longrightarrow 100$

We'll label the equal-factor pair at the upper right corner of each square. Now we will draw the diagonal line connecting the upper right-hand corners. This particular diagonal line is called the *line of squares.*

EXPECTED RESPONSE

Some students may need further convincing that squares are rectangles.

TEACHER ACTION	TEACHER TALK	EXPECTED RESPONSE
Illustrate on the grid for 12 using a colored marker.	Let's draw the line of squares on the grid for 12. Label the points like this.	
	What is the relationship between the line of squares and the cutoff?	The line of squares goes through the cutoff: it is the cutoff; the rectangles are just reversed on either side.
	Notice that the line of squares cuts exactly between the factor pairs where the reverse occurs. With 12, the line of squares goes between 3 × 4 and 4 × 3. If I label the corner of the rectangles (3,4) and (4,3) and connect these with a dotted line, the dotted line cuts the line of squares between (3,3) and (4,4).	
	Note: The graph of whole number factor pairs of a number like 22 will not have enough points to give a true picture of the graph of all numbers whose product is 22. Be careful of applying the graphic approach in such cases.	
	The line of squares locates the cutoff. If we find where the line of squares is crossed, we'll have the cutoff. For 12, we crossed the line of squares between 3 × 3 and 4 × 4. This means that we will find no new factor pairs after we check 3.	

TEACHER ACTION	TEACHER TALK	EXPECTED RESPONSE
Draw the line of squares on the other grids and repeat the illustration.		
Be sure to point out that the cutoff for 36 occurs *at* 6 × 6, which is *on* the line of squares. 36 is a square number.	Where was the line of squares crossed for the examples on your sheet?	For 12, between 3 × 3 and 4 × 4. For 20, between 4 × 4 and 5 × 5. For 36, at 6 × 6. For 30, between 5 × 5 and 6 × 6. For 40, between 6 × 6 and 7 × 7.

TEACHER ACTION	TEACHER TALK	EXPECTED RESPONSE
	So if we are looking for all the factor pairs of 40, we find no new ones after 5 × 8.	
	Let's try a number without using a grid.	
	How can we find the cutoff for 52?	Locate 52 between two square numbers.
	What two squares is 52 between?	Between 49 and 64.
	Where will we cross the line of squares for 52?	52 is between 7 × 7 and 8 × 8. So the cutoff is between 7 and 8.
	There are *no* new factor pairs to be found after we've tested through 7.	
	Find the factor pairs for 52.	1 × 52 2 × 26 4 × 13 5 No 6 No 7 No
	Not all the numbers we test will be part of a factor pair. However, we will be sure that we have found all the factor pairs when we check up to the cutoff line.	
	How far must we test to find the factor pairs for 71?	71 is between 8 × 8 and 9 × 9 so we must check through 8.
	Test them.	1 × 71 2 No 3 No 4 No 5 No 6 No 7 No 8 No
	What factors did we find?	1 and 71.
	What kind of numbers have just two factors?	Prime numbers.

TEACHER ACTION	TEACHER TALK	EXPECTED RESPONSE
Ask.	If we are looking for the factor pairs for 91, how far must we check?	91 is between 9 × 9 and 10 × 10.
	Where is the cutoff?	Between 9 and 10.
You may want to review divisibility rules.	What are the possible factors to check?	1 2, 3, 4, 5, 6, 7, 8, 9.
91 is odd and not divisible by 5. That leaves 3, 7 and 9 to be checked. You may want to relate the cutoff to the square root of the number. On a calculator press 91 [√]. The whole-number part of the square root is the largest number to be tested.	Check them.	1 × 91 7 × 13 ——— 13 × 7 91 × 1
	Let's try 49. What is the cutoff? Square numbers are easy. The cutoff is at the equal-factor pair. For numbers that are not squares, we locate the number between two square numbers to find the cutoff.	At 7 × 7.
	To find all the factor pairs for a number, I only need to check all the numbers through 15. What could my number be? Why?	15 × 15 = 225 and 16 × 16 = 256, so all the numbers from 225 up to 255 will have a cutoff between 15 and 16. That means that any number from 225 to 255 will work.
Distribute Worksheet 3-2, Crossing the Line, and Worksheet 3-3, Practice Problems. Worksheet 3-3 may need more examples. Allow students time to share their ideas and strategies about the problems.	Here are some problems to provide practice. You will have to write a longer list of square numbers and their equal-factor pairs to answer all the questions.	

NAME ____________________

Factor Pairs and Rectangles

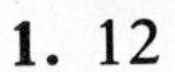

1. 12 **Pairs**

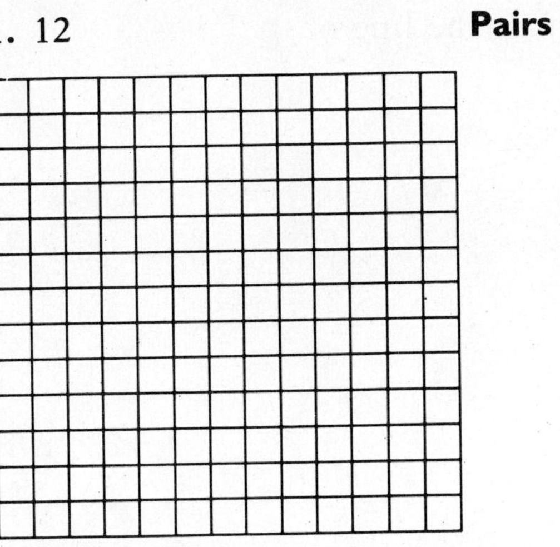

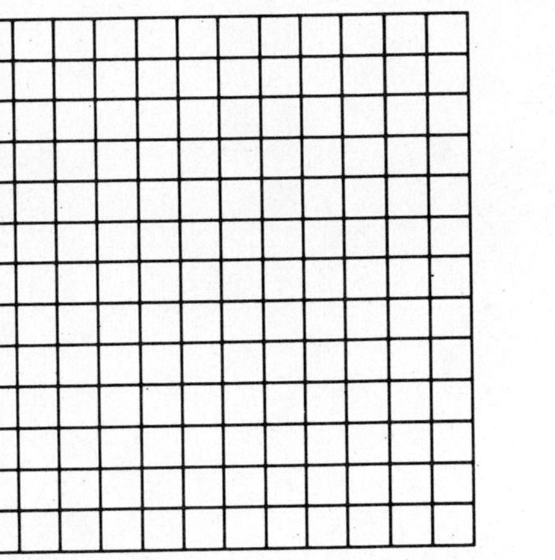

2. 20 **Pairs**

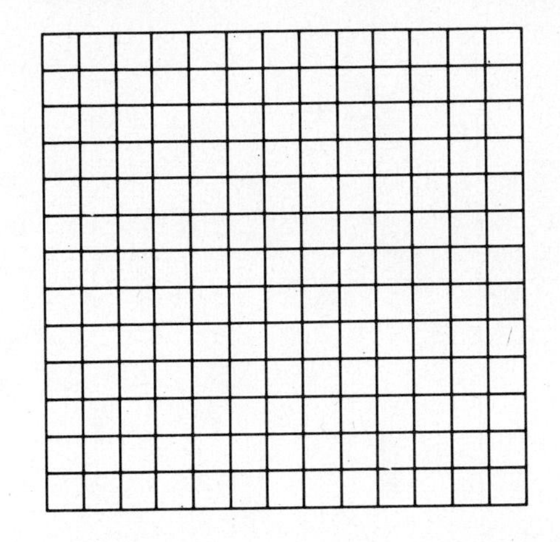

3. 36 **Pairs**

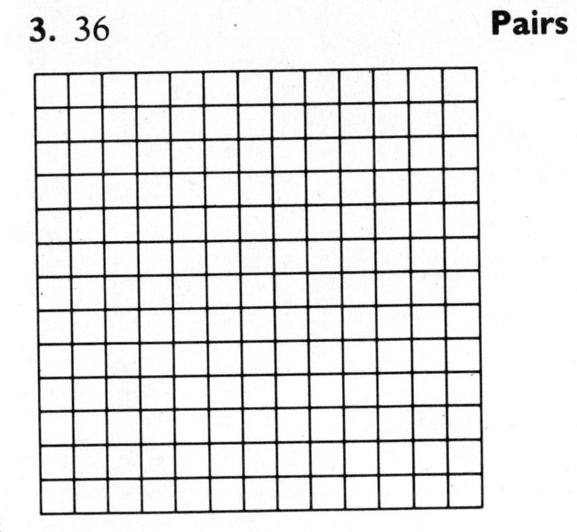

4. 30 **Pairs**

5. 40 **Pairs**

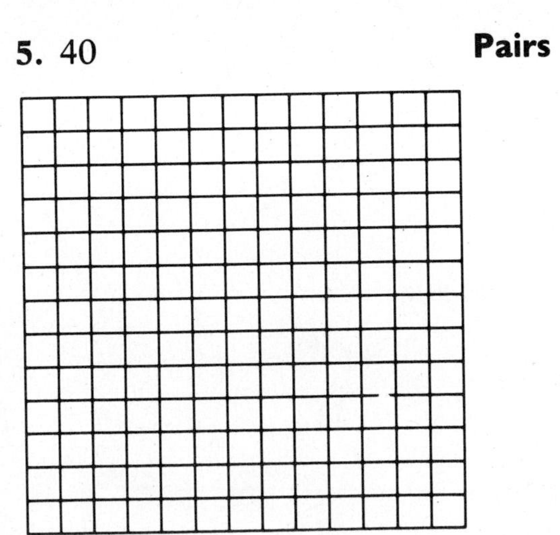

6. Squares

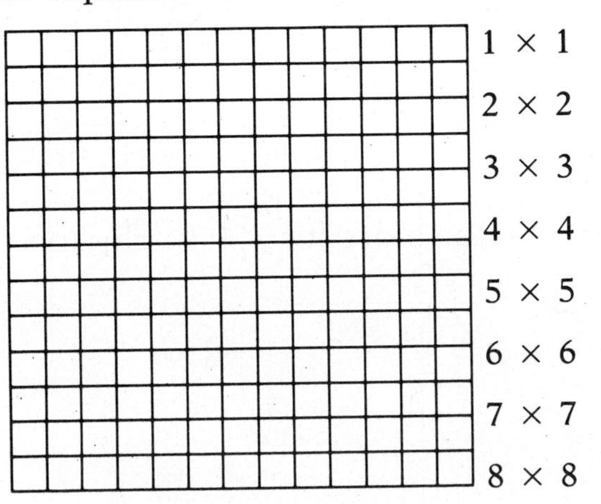

Draw the squares

1×1

2×2

3×3

4×4

5×5

6×6

7×7

8×8

NAME ______________________

Crossing the Line

For each of the numbers, find where the factor pairs cross the line of squares.

1. 22

2. 44

3. 64

4. 128

5. 154

6. 278

7. To find all the factor pairs for 300, how far do you have to check to be sure that you have found them all?

8. List all the factor pairs for 300 (including the reversed ones).

NAME ____________________

Practice Problems

1. Mr. Brown wrote a number on the blackboard and said, "I know that to be sure I find all the factor pairs for this number, I must check each number from 1 to 47."

a) What is the smallest number Mr. Brown could have written on the blackboard?

b) What is the largest number Mr. Brown could have written on the blackboard?

2. Bob is making factor pairs of a number. He finds he can make exactly 7 factor pairs (including the reversed ones). If his number is less than 100, what is it?

3. Jo has chosen a number larger than 12 and less than 50. She finds exactly 6 factor pairs (including the reversed ones). What could her number be?

Activity 4 FACTOR TREES

OVERVIEW

The Prime Factorization Theorem is of fundamental importance to a study of arithmetic. It states that every whole number can be factored into a product of prime numbers in exactly one way. It provides a basic structure for whole numbers and the relationships among whole numbers. The ideas studied in the first three activities—factor, divisor, multiple, composite, and prime—have laid the groundwork for the discovery of the notion of a prime factorization and its uniqueness.

Activity 4 starts with the Product Puzzle in which students find different strings of numbers such that the products of each string of numbers is 1,350. In finding the strings, students use the strategy of finding factor pairs. Working on the puzzle leads to two conclusions: There are many ways to write a number, such as 1,350, as a product, but there is only one "longest" way. Factor Trees provide a technique for finding prime factorization, and the Multiplication Mazes give a fun payoff for the process.

Goals for students

1. Recognize the multiplicity of factorizations of composite numbers.
2. Given one factorization of a composite number, find another.
3. Recognize the uniqueness of prime factorization.
4. Find the prime factorization of a given number.

Materials

Calculators.

Worksheets

*4-1, Product Puzzle I.
4-2, Factor Trees I.
4-3, Factor Trees II.
4-4, Multiplication Mazes.
4-5, Practice Problems.
4-6, Product Puzzle II.

Transparencies

Starred item should be made into a transparency.

FACTOR TREES

TEACHER ACTION	TEACHER TALK	EXPECTED RESPONSE
Students need calculators and Worksheet 4-1, Product Puzzle I.		
Put the following factor pairs on the overhead. Reveal one pair at a time.	Multiply these numbers on your calculators.	
15×28	What do you find?	They all equal 420.
$20 \times 7 \times 3$ $3 \times 35 \times 4$	There are many ways to write 420 as a product. Can anyone find another way?	Students will suggest other ways.
Take three or four responses, checking each one.	How did you find that one?	Elicit, if possible, the *combining* strategy: I took $20 \times 7 \times 3$ and multiplied $7 \times 3 = 21$ and got 20×21 and the *breaking apart strategy* I took 15×28 and I wrote 3×5 instead of 15, so $3 \times 5 \times 28$ is a solution.
Display a transparency of Worksheet 4-1, Product Puzzle I, on the overhead.	The Product Puzzle is like a word search puzzle except that we are going to group together strings of numbers whose product is 1,350. The strings can go horizontally, vertically, or around corners. Let's find one together.	
Circle ($5 \times 6 \times 5 \times 9$) $\times$ 54; check the product on a calculator.	Look at the top row. Can you find a string of numbers whose product is 1,350?	Yes! $5 \times 6 \times 5 \times 9$
Circle (5×3 / $10 \times$ / $\times$ / 9) in upper left corner.	Try this one. Does it work?	Yes!
	In this puzzle there are enough lines for 15 strings. Find as many as you can and answer the two questions at the bottom of the page. You can use the strings we found together as a start.	

OBSERVATIONS	**POSSIBLE RESPONSES**
As an extra challenge, have students find all the factors of 1,350. (There are 24.)	The more advanced students will observe that they can find new strings in their heads and don't need the calculator.
1, 2, 3, 5, 6, 9, 10, 15, 18, 25, 27, 30, 45, 50, 54, 75, 90, 135, 150, 225, 270, 450, 675, 1,350.	For those unable to find any strings, say, "Look at the puzzle. See if you can find 5, 6, 5, and 9 again."
Pairs (1, 1,350) (2, 675) (3, 450) (5, 270) (6, 225) (9, 150) (10, 135) (15, 90) (18, 75) (25, 54) (27, 50) (30, 45)	For those unable to find different strings, say, "You know 5 × 5 × 6 × 9 = 1,350. Can you put the two 5s together? Can you break anything apart?" etc.

TEACHER ACTION	TEACHER TALK	EXPECTED RESPONSE
Lead the students in analyzing how they looked for strings.	How did you go about finding strings? What were you thinking about? You multiplied together some of the factors of a string you already had. And you broke apart factors.	Elicit responses such as Combine: $5 \times 5 \times 9 \times 6 = 1{,}350$ so $25 \times 9 \times 6 = 1{,}350$. Break Apart: $5 \times 3 \times 10 \times 9 = 1{,}350$ so $5 \times 3 \times 5 \times 2 \times 9 = 1{,}350$.
	Are there any other ways you looked at the problem?	I broke factors apart, then put them together differently. Occasionally, a student will use division to find other factors: $1{,}350 \div 45 = 30$.
Make a list of new and different strings on the board or overhead. If a new string has a 1, discuss whether that's really different. Be sure the point is made that a string could be multiplied by lots and lots of 1s, but the new string wouldn't have any different factors. When we ask for the longest string, we will not include 1s.	What new and different strings did you find for question 1? How did you find them?	Various answers.
	What was the longest string you found for question 2?	Various answers; it may take awhile before students realize they are all the same. $2 \times 3 \times 3 \times 3 \times 5 \times 5$
If you have not discussed the repeated factors of 1, do it now.	Can you break this down further?	No, the factors are primes.

Activity 4 *Summarize*

TEACHER ACTION	TEACHER TALK	EXPECTED RESPONSE
Break down two or three more of the strings on the board.	Let's take an example of a string for 1,350 from your list. We want to break down each factor until we have *all primes*.	
Example: $3 \times 25 \times 9 \times 2$ gives $3 \times 5 \times 5 \times 3 \times 3 \times 2$.	What do we always get?	$2 \times 3 \times 3 \times 3 \times 5 \times 5$ in some order.
Put $15 \times 28 = 420$ on the board.	What is the longest string you can break 420 into?	$2 \times 2 \times 3 \times 5 \times 7$
With fifth and sixth graders you may need to have the class find a longest string for other numbers before proceeding.	What are all the factors in a longest string?	Prime numbers.
	What is the longest string for 24? What is it for 60? How will we know when we have found the *longest* string?	$24 = 2 \times 2 \times 2 \times 3$; $60 = 2 \times 2 \times 3 \times 5$. We will have all prime numbers.
	We call the longest string of factors we can find for a number the *prime factorization* of the number.	
	Today we have discussed that a number has only one longest string; therefore, we will say that the prime factorization of a number is *unique*. There is only one prime factorization for a number.	

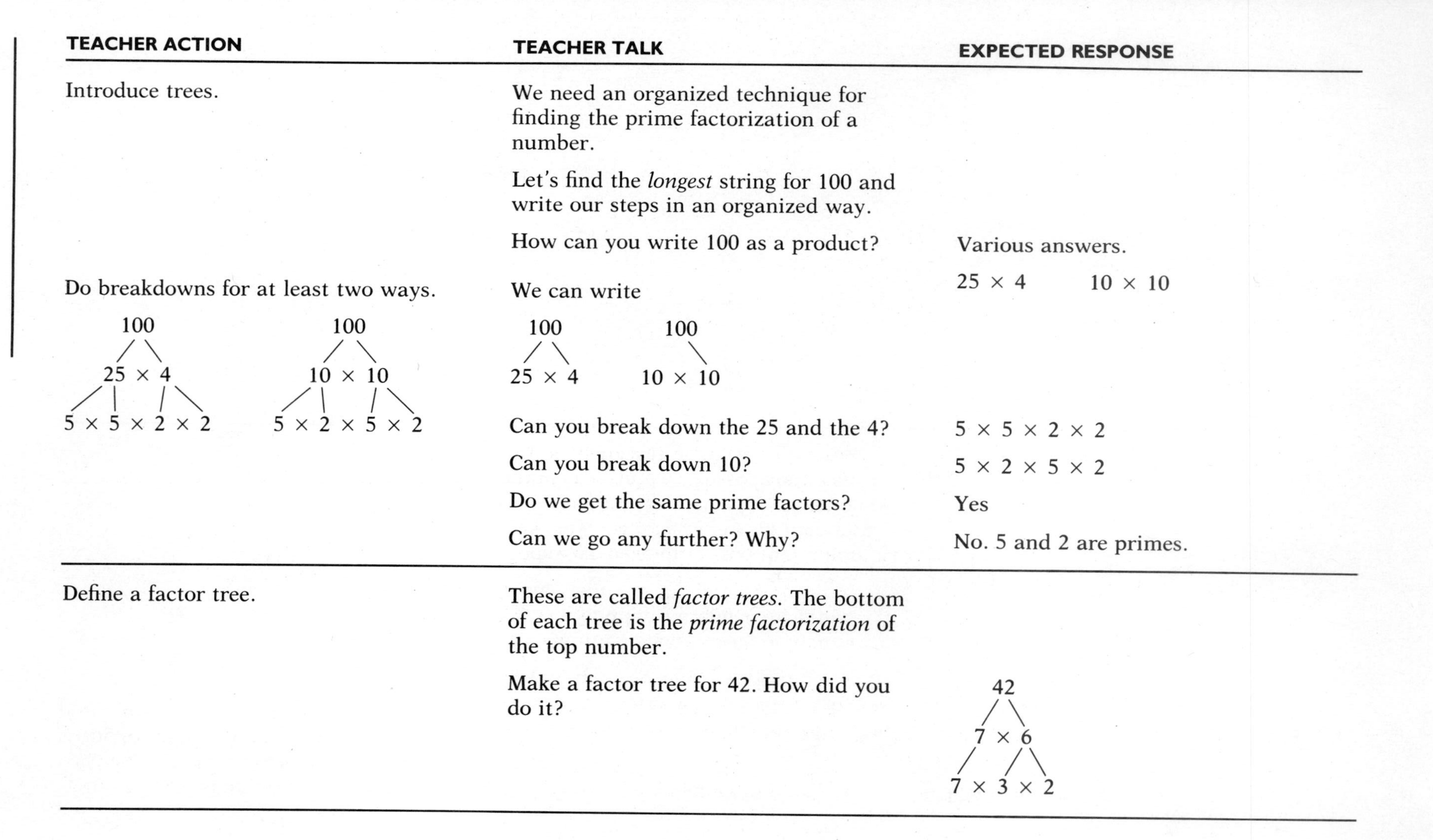

TEACHER ACTION	TEACHER TALK	EXPECTED RESPONSE
Introduce trees.	We need an organized technique for finding the prime factorization of a number.	
	Let's find the *longest* string for 100 and write our steps in an organized way.	
	How can you write 100 as a product?	Various answers.
Do breakdowns for at least two ways.	We can write	25×4 10×10
100 → 25×4 → $5 \times 5 \times 2 \times 2$ 100 → 10×10 → $5 \times 2 \times 5 \times 2$	100 → 25×4 100 → 10×10	
	Can you break down the 25 and the 4?	$5 \times 5 \times 2 \times 2$
	Can you break down 10?	$5 \times 2 \times 5 \times 2$
	Do we get the same prime factors?	Yes
	Can we go any further? Why?	No. 5 and 2 are primes.
Define a factor tree.	These are called *factor trees.* The bottom of each tree is the *prime factorization* of the top number.	
	Make a factor tree for 42. How did you do it?	42 → 7×6 → $7 \times 3 \times 2$

Activity 4 *Summarize*

TEACHER ACTION	TEACHER TALK	EXPECTED RESPONSE
Illustrate on overhead after students have tried the problem.	Find the prime factorization of 108. How could we write 108 as a product? Use your calculators if you need to.	It's an even number, so 2 is a factor. 108 2×54 $2 \times 6 \times 9$ $2 \times 3 \times 2 \times 3 \times 3$
	Make a factor tree for 153.	
	It's not an even number, so 2 isn't a factor.	
	What could we try?	Various answers. Test the answers.
Put factor tree for 153 on the board. 153 3 51 $3 \times 3 \times 17$	If 3 is a factor of 153, then $153 = 3 \times \square$	
	Here $\square$ stands for another factor.	
	How do we find the other factor?	Divide: $153 \div 3 =$ *51*.
	To review, what do we do first if we want to find the *prime factorization* of a number?	Find a pair of factors.
	What do you do if you don't recognize a pair of factors?	Various answers; check to see if the number is even or ends in 0 or 5 (if students know other divisibility tests encourage them to use them); start with 2 and check all the numbers up to the cutoff for the number.
A side trip here would be to develop divisibility tests.		
Pass out Worksheet 4-2, Factor Trees I, and Worksheet 4-3, Factor Trees II.	Factor Trees I and II give you a chance to practice the techniques we have developed. Make factor trees for the numbers given.	

TEACHER ACTION	TEACHER TALK	EXPECTED RESPONSE
After students have finished the factor trees, discuss the Multiplication Mazes. This provides a fun payoff for prime factorization.		
Draw the following mazes on the blackboard or overhead. Write: $2 \times 2 \times 2 \times 3 \times 5 \times 7$.	In the last activity you found that $840 = 2 \times 2 \times 2 \times 3 \times 5 \times 7$.	
	This is a Multiplication Maze. We must find a path of numbers from the entrance to the exit cell whose product is the number in the exit cell, 840. No diagonal moves are allowed.	
	Notice that the 2 at the entrance and the 2 at the exit must be used. So we need a path from the 2 to the other 2 that has factors 2, 3, 5, and 7.	
	It helps to write factors in the cells that contain a composite number.	
	What path is the correct one?	
	What factors will the path contain?	$2 \times 2 \times 2 \times 3 \times 5 \times 7$.
	The factors in the path must be exactly the prime factorization of the exit number.	

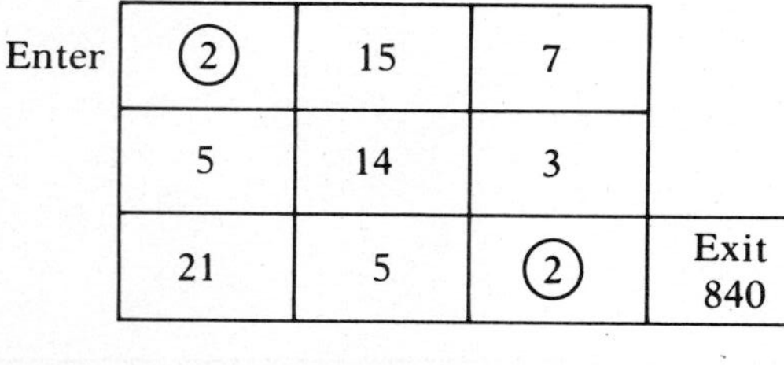

Enter	(2)	(3 × 5) 15	7	
	5	(2 × 7) 14	3	
	(3 × 7) 21	5	(2)	Exit 840

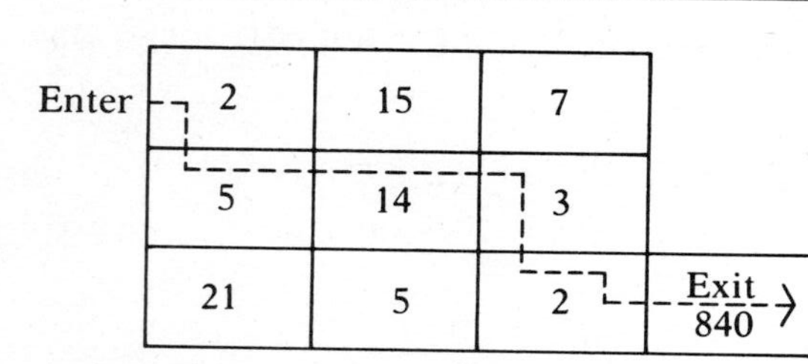

TEACHER ACTION	TEACHER TALK	EXPECTED RESPONSE
Pass out Worksheet 4-4, Multiplication Mazes.	Now find the paths in the Multiplication Mazes worksheet.	
Worksheet 4-5, Practice Problems, and Worksheet 4-6, Product Puzzle II, are included as a possible homework assignment or as an extra challenge.		

NAME ______________________

Product Puzzle I

Draw a loop around strings of numbers whose product is 1,350.

5	×	6	×	5	×	9	×	54
×	3	×	3	×	150	×	9	×
10	×	27	×	5	×	5	×	2
×	54	×	25	×	3	×	45	×
9	×	5	×	6	×	5	×	135
×	5	×	9	×	150	×	2	×
15	×	10	×	9	×	3	×	5
×	5	×	6	×	45	×	15	×
1	×	3	×	25	×	9	×	2

Record the different strings you have found.

1. ______________
2. ______________
3. ______________
4. ______________
5. ______________
6. ______________
7. ______________
8. ______________
9. ______________
10. ______________
11. ______________
12. ______________
13. ______________
14. ______________
15. ______________

After you have found 15 different strings, answer these questions.

1. Find 2 strings of numbers whose product is 1,350 that are not in the table.

2. Can you find a string whose product is 1,350 that is longer than any in the table?

NAME

Factor Trees I

For each of these numbers, grow a factor tree. When you reach the bottom of a tree, check it by multiplying the numbers together.

1. 36	**2.** 84
3. 72	**4.** 57
5. 144	**6.** 147
7. 63	**8.** 64

NAME

Factor Trees II

For each of these numbers, grow a factor tree. When you reach the bottom of a tree, check by multiplying the numbers together.

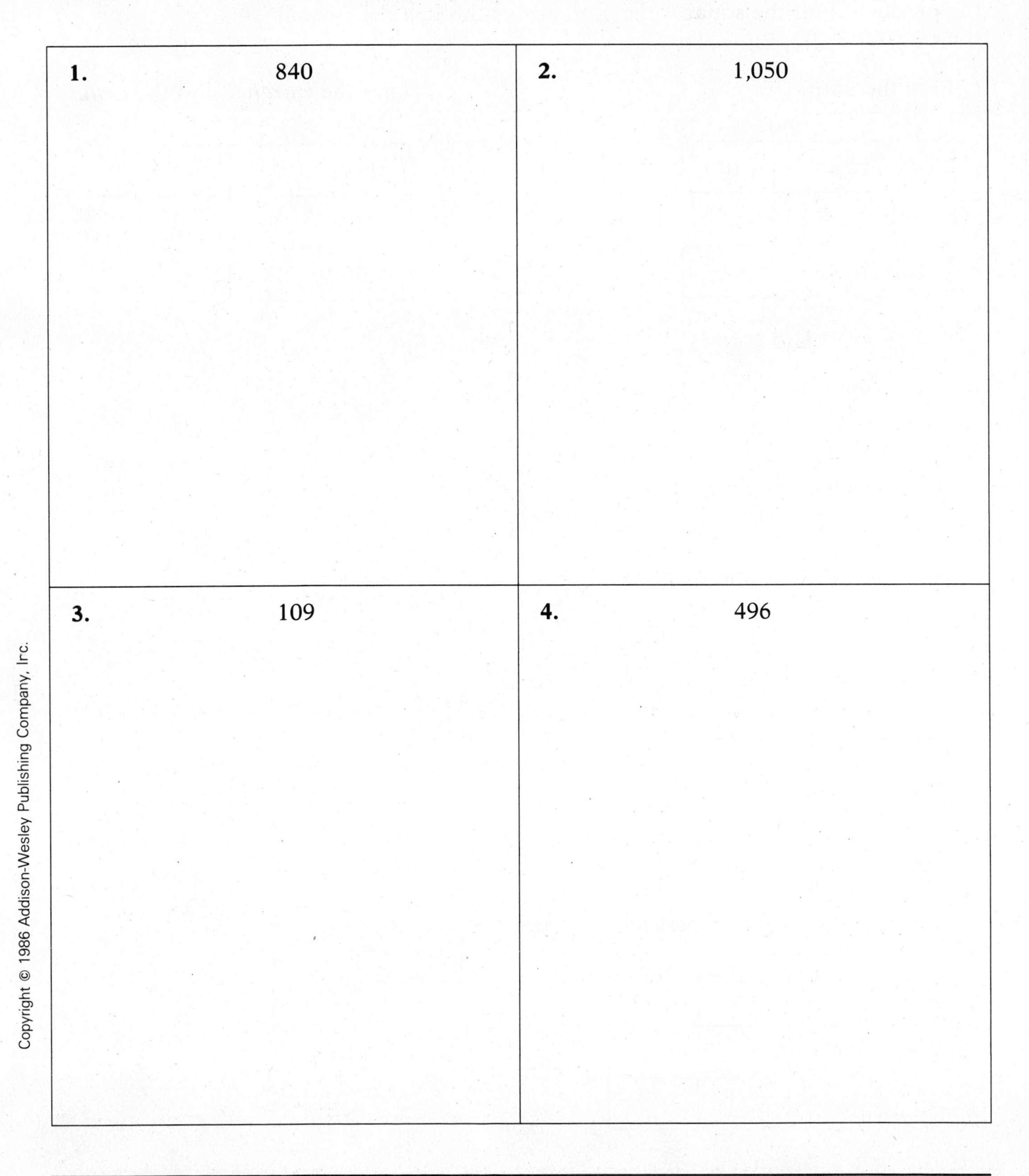

NAME

Multiplication Mazes

How can you use the fact that $840 = 2 \times 2 \times 2 \times 3 \times 5 \times 7$ to find the path through these mazes? No diagonals! Remember that the product of all the squares the path passes through must equal the exit number.

1. Find the path.

enter

5	3	10
6	7	2
11	2	4
	840 exit	

2. Find the *entrance* and the *path.*

14	2	2	
3	3	2	exit 840
5	7	6	

3. Make a factor tree for 3,927:

3,927

Find the entrance and the path.

8	5	11	
2	13	3	exit 3,927
11	17	7	

4. Make up your own numbers for this maze.

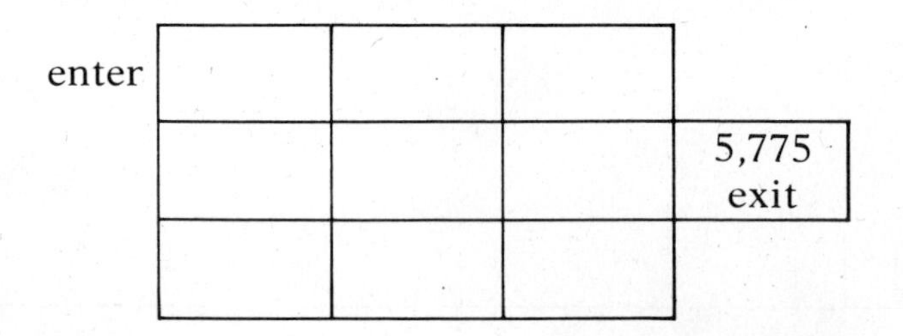

NAME

Practice Problems

1. Write the last four digits of your telephone number in order.

2. Find the prime factorization of the number you wrote in problem 1.

3. How many different prime factors are in your number?

4. How many prime factors greater than 13 are in your number?

5. Is your number prime?

6. Is your number divisible by 6?

7. Is your number divisible by 14?

8. Is your number divisible by 10?

9. Is your number divisible by 15?

NAME ____________________

Product Puzzle II

Find all strings whose product is 630. Strings may go horizontally, vertically, or bend around corners.

7	×	3	×	30
×	5	×	42	×
6	×	15	×	21
×	45	×	9	×
5	×	14	×	2

Activity 5 COMMON MULTIPLES

OVERVIEW

Common multiples of numbers can be found in two ways. One way is to use the calculator to generate the multiples of each number and then examine each sequence of multiples to generate a sequence of common multiples and find the first (smallest) multiple, or least common multiple (LCM). The second method of finding the LCM is to build the prime factorization for the LCM from the prime factorization of each number.

Goals for students

1. Learn the definition of *common multiple, least common multiple,* and *relatively prime.*
2. Find common multiples and least common multiples by listing.
3. Use prime factorization to find least common multiple.
4. Use the concept of common multiple and least common multiple to solve problems.

Materials

Calculators.

Worksheets

5-1, Least Common Multiples.

5-2, Practice Problems.

5-3, Applications of LCM.

COMMON MULTIPLES

TEACHER ACTION	TEACHER TALK	EXPECTED RESPONSE
Play a guessing game.	I'm thinking of a number; 3 is a factor of my number and 4 is a factor of my number.	
	What number could I have in mind?	12; 24; and so on.
	How many other numbers would work?	Many, lots, more than we can count, . . . , an infinite number.
If students cannot guess any of the correct numbers, ask.	What numbers have 3 as a factor?	3, 6, 9, 12, 15, 18, 21, 24, and so on.
	What do we call these numbers?	Multiples of 3.
	What numbers have 4 as a factor?	4, 8, 12, 16, 20, 24, 28, and so on.
	What do we call these numbers?	Multiples of 4.
	Look at both sets of numbers.	
	Which numbers occur in both sets?	12, 24, 36, and so on.
	What do we call these numbers?	Numbers that are multiples of both 3 and 4.
Explain.	We call these numbers *common multiples* of 3 and 4.	
Ask.	How can we tell if a number is a common multiple of 3 and 4?	Check to see if 3 is a factor and check to see if 4 is a factor.
	Is 48 a common multiple of 3 and 4? Why?	Yes; $48 = 4 \times 12$ and $48 = 3 \times 16$.
	Is 92 a common multiple of 3 and 4? Why?	No; $92 \neq 3 \times$ (a number); 3 is not a factor of 92.

Activity 5 *Launch*

TEACHER ACTION	TEACHER TALK	EXPECTED RESPONSE
Play the guessing game again.	I'm thinking of another number. 6 is a factor of my number and 9 is a factor of my number. My number is more than 50 but less than 100. What is it?	72, 54, or 90.
	How did you find it?	Find all the common multiples of 6 and 9 up to 100. Look for the one that is between 50 and 100.
	How did you find the common multiples?	Use a calculator to find all the multiples of 6 and all the multiples of 9 and then look for the common multiple.
If students are having difficulty, use more examples to illustrate the method of finding multiples.		
Play the guessing game again.	I'm thinking of the *smallest* number that has both 3 *and* 4 as factors. What could the number be?	12
	Would any other numbers work?	No.
	We call 12 the *smallest* or *least common multiple* (LCM) of 3 and 4.	
Define LCM.	The smallest common multiple or first common multiple of two numbers is called the *least common multiple* (LCM).	
Explain and ask.	12 is the LCM of 3 and 4 and $12 = 3 \times 4$. Do you think the LCM is always the product of the two numbers?	Various answers. If students don't know, tell them we will explore more possibilities and answer the question later.

TEACHER ACTION	TEACHER TALK	EXPECTED RESPONSE
With the entire class, use calculators to practice finding the LCM's. Write results on the board or overhead.		
Try 6 and 8.	Find the LCM of 6 and 8 and show why it is the LCM.	Multiples of 6: 6, 12, 18, 24, 30, 36 . . . Multiples of 8: 8, 16, 24, 32, 40 . . . Common multiples of 6 and 8: 24, 48, 72 . . .
	What can you say about the common multiples of 6 and 8?	They increase by 24 each time; they are multiples of 24.
	What is the LCM of 6 and 8?	LCM of 6 and 8: 24 $24 = 6 \times 4$ $24 = 8 \times 3$
Try 4 and 15. Repeat leading questions.	What is the LCM of 4 and 15?	LCM of 4 and 15: 60 $60 = 4 \times 15$
Try 12 and 15. Repeat leading questions.	What is the LCM of 12 and 15?	LCM of 12 and 15: 60 $60 = 12 \times 5$ $60 = 15 \times 4$
	How do you know?	
If necessary try more examples before going on to the factorization method of finding LCM.		
Ask.	Now do you think that LCM is the product of the two numbers? Why?	No; the LCM of 6 and 8 $\neq$ 48 and the LCM of 12 and 15 $\neq$ 180.
	Is the product of two numbers always a common multiple of the two numbers? Why?	Yes; each number is a factor of the product.
	Let's see if we can find a way to predict when the LCM is the product of the two numbers. To do this we will examine another way to find LCM.	

TEACHER ACTION	TEACHER TALK	EXPECTED RESPONSE
Look at factorizations.	What was the LCM of 3 and 4?	12.
	How is 12 related to 3 and 4?	12 is a multiple of 3 and 4. 3 and 4 are both factors of 12.
We can show that 3 and 4 are factors of 12 by looking at factorizations.	Make a factor tree for 3, 4, and 12.	(3) → 3; (4) → 2 × 2; (12) → 2 × 2 × 3
	How does the factorization of 12 compare to the factorizations of 3 and 4?	The prime factorization of 12 contains the prime factorizations of both 3 and 4. (2 × 2) × (3), with (2 × 2) → 4 and (3) → 3
Ask.	What was the LCM of 12 and 15?	60.
	Make a factor tree for 12, 15, and 60.	(12) → 2 × 2 × 3; (15) → 3 × 5; (60) → 2 × 2 × 3 × 5
	Does the factorization of 60 contain the factorization of 12 and 15?	60 = (2 × 2 × (3) × 5), with 2 × 2 × 3 → 12 and 3 × 5 → 15
	Show me by circling the factors.	

TEACHER ACTION	TEACHER TALK	EXPECTED RESPONSE
Ask.	Can you describe the prime factorization of the LCM?	Various answers. Elicit that the prime factorization is the shortest string that has both the other strings in it.
	For the LCM of 12 and 15 we need a string with both $2 \times 2 \times 3$ and 5×3 in it. If we start with $2 \times 2 \times 3$, what else do we need?	5.
Display on the overhead and circle in two colors.	So the string is $2 \times 2 \times 3 \times 5$. Let's circle each number 12 and 15 to be sure we have both as a factor of this string.	

(2 × 2 × (3) × 5)

12   15

For students having difficulty, display the factorizations this way:

$$12 = 2 \times 2 \times 3$$
$$15 = 3 \times 5$$
$$\text{LCM} = 2 \times 2 \times 3 \times 5$$

$$24 = 2 \times 2 \times 2 \times 3$$
$$42 = 2 \times 3 \times 7$$
$$\text{LCM} = 2 \times 2 \times 2 \times 3 \times 7$$

Try more examples if necessary:
4 and 21, 12 and 24, 30 and 42.

TEACHER ACTION	TEACHER TALK	EXPECTED RESPONSE
Ask.	What is the LCM of 4 and 15? Why?	60 4 → 2×2; 15 → 3×5 $2 \times 2 \times 3 \times 5 = 60$
	In this case the LCM of 4 and 15 is 4×15.	
Ask.	When is the product of the two numbers the LCM?	When the two numbers have no factor in common. If students don't understand, practice with a few more pairs of numbers, such as 9 and 16 or 9 and 15.
	When two numbers have no factors in common we say that they are *relatively prime.* So the LCM is the product of the two numbers when the numbers are *relatively prime.*	
	Let's go the other way.	
	What is the LCM of the numbers $2 \times 2 \times 3 \times 5$ and $3 \times 3 \times 5 \times 7$?	
	Let's figure out the shortest string to contain both of these.	
	It certainly must have $2 \times 2 \times 3 \times 5$.	
	What else does it need?	Another 3 and a 7.
	What is the LCM?	$2 \times 2 \times 3 \times 5 \times 3 \times 7$
Circle both numbers to check $2 \times 2 \times 3 \times 5 \times 3 \times 7$ (with $2 \times 2 \times 3 \times 5$ circled and $3 \times 5 \times 3 \times 7$ circled)	What have we just found?	The LCM of 60 and 315 is 1,260.
Pass out Worksheet 5-1, Least Common Multiples. Assign Worksheet 5-2, Practice Problems, and Worksheet 5-3, Applications of LCM, as homework if necessary.		

Activity 5 *Summarize*

TEACHER ACTION	TEACHER TALK	EXPECTED RESPONSE
Go over Worksheet 5-1 and 5-2.	What is the LCM of 6 and 8?	24
	What do we know about the common multiples of 6 and 8?	They are multiples of 24.
	Is the product of two numbers always a common multiple of the two?	Yes.
Check several examples with small numbers, including the pairs done so far.		
Ask.	Find the LCM of 15 and 24.	
	How did you do it?	Used calculators or prime factorization.
	What is the prime factorization of the LCM?	$2 \times 2 \times 2 \times 3 \times 5 = 120$
Try more examples if necessary.	Use the prime factorization of 120 to find some factors of 120 other than 15 and 20.	Many answers are possible: $2 \times 2 = 4$ $2 \times 3 = 6$ $2 \times 3 \times 5 = 30$ and so on.
Refer to Worksheet 5-1, problem 4, and Worksheet 5-2, problems 1, 4, and 11, if students cannot answer.	When is the LCM of two numbers equal to the product of the two numbers?	When the two numbers have no common factors.

TEACHER ACTION	TEACHER TALK	EXPECTED RESPONSE
Optional questions.	What is a common multiple of 7 and 1,001?	$7 \times 1{,}001 = 7{,}007$
	We need a number with 7 as a factor and 1,001 as a factor.	
	Could the LCM of 7 and 1,001 be more than 7,007?	No. Either 7,007 is the LCM *or* the LCM is a factor of 7,007.
	Can you make a rule about the biggest the LCM of two numbers can be?	No bigger than the product.
	How can you tell when the LCM of two numbers will be smaller than the product?	When they have a *common factor.*
	What is the LCM of 7 and 1,001?	$7 \times 11 \times 13$.
Here is another guessing game.	I have a bag of tiles. I can divide the tiles into 3 equal stacks.	
	I can divide the tiles into 4 equal stacks.	
	I have more than 90 but fewer than 100 tiles.	
Explain that $96 = 3 \times 32$ and $96 = 4 \times 24$;	How many tiles are in the bag?	96.
96 is a common multiple of 3 and 4, which is between 90 and 100.		

NAME ______________________

Least Common Multiples

In problems 1–3, use a calculator to find the multiples of each number. Write the least common multiple in the box. Show what each of the numbers must be multiplied by to get the LCM.

1. Multiples of 3: ____ ____ ____ ____ ____ ____ ____ ____

Multiples of 4: ____ ____ ____ ____ ____ ____ ____ ____

Common multiples of 3 and 4. ______________________

3 × ____
= LCM = ☐
4 × ____

2. Multiples of 30: ____ ____ ____ ____ ____ ____ ____ ____

Multiples of 42: ____ ____ ____ ____ ____ ____ ____ ____

Common multiples of 30 and 42: ______________________

30 × ____
= LCM = ☐
42 × ____

3. Multiples of 18: ____ ____ ____ ____ ____ ____ ____ ____

Multiples of 36: ____ ____ ____ ____ ____ ____ ____ ____

Common multiples of 18 and 36: ______________________

18 × ____
= LCM = ☐
36 × ____

In problems 4–6, make a factor tree for each of the numbers. Use the prime factorizations to find the least common multiple.

4. 3 4 String for LCM ☐ = ____

5. 30 40 String for LCM ☐ = ____

6. 18 35 String for LCM ☐ = ____

NAME ______________________

Practice Problems

Find the least common multiple (LCM) of each pair of numbers given.

1. 4, 9

LCM is ______

2. 8, 14

LCM is ______

3. 10, 45

LCM is ______

4. 14, 15

LCM is ______

5. 14, 21

LCM is ______

6. 24, 36

LCM is ______

7. 58, 96

LCM is ______

8. 180, 210

LCM is ______

Find all the common multiples less than 100 for each pair of numbers given.

9. Common multiples of 5 and 7 ______________________

10. Common multiples of 12 and 20 ______________________

11. Common multiples of 4 and 6 ______________________

12. Common multiples of 9 and 15 ______________________

What is the smallest number that has the following factors?

13. 2 and 3.

14. 2, 3 and 4.

15. 4, 6 and 9.

16. 4, 6, 9 and 12.

17. I am thinking of a number. The least common multiple of my number and 9 is 45. What could my number be?

18. I am thinking of a number. My number has 8 as a factor and 12 as a factor.

a) What is the smallest that my number could be?

b) Name four other numbers that are factors of my number.

19. a) Find the LCM of 12 and 35.

b) Name 3 other common multiples of 12 and 35.

c) Name 3 other factors of the LCM.

NAME ______________________

Applications of LCM

1. Gleamy-Tooth toothpaste comes in 2 sizes.

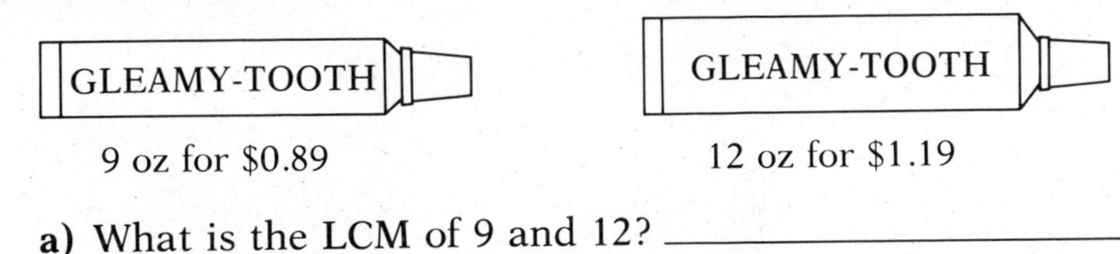

9 oz for \$0.89 12 oz for \$1.19

a) What is the LCM of 9 and 12? ______________________

b) If you bought that much toothpaste in 9-oz tubes, how much would it cost? ______________________

c) If you bought that much toothpaste in 12-oz tubes, how much would it cost? ______________________

d) Which tube gives you more Gleamy-Tooth for the money?

____

2. In the school kitchen during lunch, the timer for pizza buzzes every 14 minutes; the timer for hamburger buns buzzes every 6 minutes. The two timers just buzzed together. In how many minutes will they buzz together again?

3. Two ships sail steadily between New York and London. One ship takes 12 days to make a round trip; the other takes 15 days. If they are both in New York today, in how many days will they both be in New York again?

4. The high school lunch menu repeats every 20 days; the elementary school menu repeats every 15 days. Both schools are serving sloppy joes today. In how many days will they both serve sloppy joes again?

5. Two neon signs are turned on at the same time. One blinks every 4 seconds; the other blinks every 6 seconds. How many times per minute do they blink on together?

6. How many teeth should be on gear A if each turn of gear A is to produce a whole number of turns of the shafts attached to B and C?

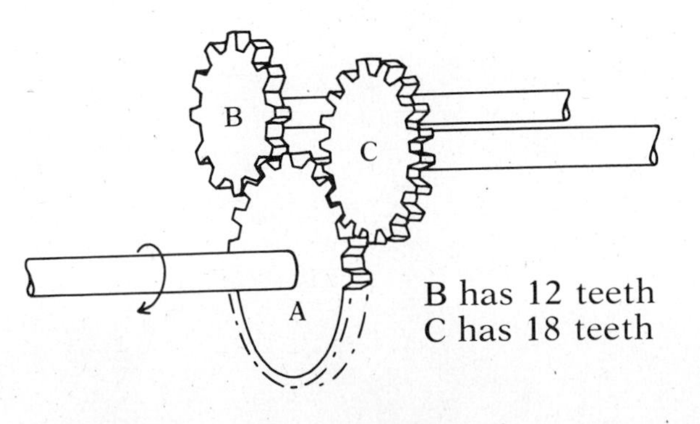

Activity 6 COMMON FACTORS

OVERVIEW

Finding factors that are common to two or more numbers or expressions is a basic simplification technique in mathematics. Students usually encounter this type of simplification first in their study of equivalent fractions; it also pervades the study and handling of algebraic expressions. In this activity the students develop a basic understanding of a common factor and the greatest common factor. They develop two techniques for finding the greatest common factor: listing all factors of both numbers and comparing, and finding the intersection of the prime factorizations of the two numbers.

Students are given further opportunity to develop their problem-solving skills through solving word problems of two types: separating a given set of objects into equal parts or distributing an equal number of objects among a given number of participants.

Goals for students

1. Learn the definition of common factor and greatest common factor.
2. Find common factors and greatest common factors by listing factors.
3. Use prime factorizations to find common factors and greatest common factors.
4. Use the concept of common factor or greatest common factor to solve problems.

Materials

Calculators.

Worksheets

6-1, Common Factors.

6-2, Greatest Common Factor.

6-3, Practice Problems.

COMMON FACTORS

TEACHER ACTION	TEACHER TALK	EXPECTED RESPONSE
Students need calculators.	What are the factors of 36? To make the work easier, remember factor pairs.	1, 2, 3, 4, 6, 9, 12, 18, 36
Record the factors of 36 and 48.	What are the factors of 48?	1, 2, 3, 4, 6, 8, 12, 16, 24, 48
Indicate the factor pairs.	What numbers appear on both lists?	1, 2, 3, 4, 6, 12
Define *common factor*.	These are the *common factors* of 36 and 48. They are factors of both 36 and 48.	
	What are the factors of 30?	1, 2, 3, 5, 6, 10, 15, 30
	What are the factors of 18?	1, 2, 3, 6, 9, 18
	What are the common factors?	1, 2, 3, 6
	What are the common factors of 18 and 9?	1, 3, 9
	What are the common factors of 18 and 25?	1
	What number will always appear when we make a list of common factors?	1
Review definition of *relatively prime*.	If the only common factor of two numbers is 1, we say that the numbers are *relatively prime*. This means that they have no common factors other than 1.	
Ask.	I'm thinking of two numbers that have no common factors other than 1. They are both less than 20. One number is 12. Could the other number be 2?	No; 2 and 12 have 2 as a common factor.
Check each response.	What could the other number be?	1, 5, 7, 11, 13, 17, 19
	Same game. I'm thinking of two numbers less than 20. They are relatively prime. One number is 9.	

TEACHER ACTION	TEACHER TALK	EXPECTED RESPONSE
Give another example.	What could the other number be?	1, 2, 4, 5, 7, 8, 10, 11, 13, 14, 16, 17, 19
	What must be true of the numbers in the list?	They must not have 3 or 9 as a factor.
Ask.	What are the common factors of 144 and 26?	
	How shall we find out?	Make a list.
	What are the factors of 144?	1, 2, 3, 4, 6, 8, 9, 12, 16, 18, 24, 36, 48, 72, 144
	What are the factors of 26?	1, 2, 13, 26
	What are the common factors?	1, 2
	How could we have found the common factors without writing that long list for 144?	Write the short list for 26 and test each number.
	Good. Common factors are on both lists. If we test every number on the shorter list, we'll find all the common factors.	
	How do we test whether 26 is a factor of 144?	See if 144 ÷ 26 comes out even.
	Try that strategy to find the common factors of 3,003 and 26.	Factors of 26 are 1, 2, 13, 26. Test each one; 1 and 13 can be divided into 3,003.

Activity 6 *Launch*

TEACHER ACTION	TEACHER TALK	EXPECTED RESPONSE
	Coach Jones is bringing 24 pretzels and 36 cookies for the mudball team. These have been carefully counted so that each team member gets the same number of pretzels and each team member gets the same number of cookies.	
Ask for a reason for each answer suggested. Ask how many pretzels and cookies each player gets.	How many team members are there?	Various answers; 1, 2, 3, 4, 6, and 12 are correct. If the students say that you cannot tell from what's given, ask for a list of all numbers that could be the answer.
	The common factors of 36 and 24 are 1, 2, 3, 4, 6. Make a table.	
Elicit table entries from class.		(see table below)
Students need Worksheet 6-1, Common Factors.		

Players	Pretzels	Cookies
1	24	36
2	12	18
3	8	12
4	6	9
6	4	6

Activity 6 *Summarize*

TEACHER ACTION	TEACHER TALK	EXPECTED RESPONSE
Review answers to problems 1 and 2 on Worksheet 6-1, Common Factors.	For 9 and 24, what are the common factors?	1, 3
	What was the (largest) *greatest common factor?*	3
	What are the common factors of 32 and 48?	1, 2, 4, 8, 16
	What is the greatest common factor?	16
Define.	The greatest common factor (GCF) of two numbers is the largest number that is a factor of both numbers.	
Record.	List the prime factorization of 9 and of 24.	$9 = 3 \times 3$
$9 = 3 \times 3$		$24 = 2 \times 2 \times 2 \times 3$
$24 = 2 \times 2 \times 2 \times 3$	What are the factors that occur in both prime factorizations?	Only the 3.
Circle.		
$9 = 3 \times (3)$		
$24 = 2 \times 2 \times 2 \times (3)$		

TEACHER ACTION	TEACHER TALK	EXPECTED RESPONSE
Record. $32 = \boxed{2 \times 2 \times 2 \times 2} \times 2$ $48 = \boxed{2 \times 2 \times 2 \times 2} \times 3$	What are the prime factorizations of 32 and 48?	$2 \times 2 \times 2 \times 2 \times 2$ $2 \times 2 \times 2 \times 2 \times 3$
	What is the largest string I can circle in both?	$2 \times 2 \times 2 \times 2$
	How much is that?	16
	16 is the largest or greatest common factor.	
Record.	What about 14 and 15?	
$14 = 2 \times 7$ $15 = 3 \times 5$	What are the prime factorizations of 14 and 15?	$14 = 2 \times 7$ $15 = 3 \times 5$
	What is the largest string that is in both factorizations?	There isn't any.
	They have *no* common prime factors. There is a common factor that isn't a prime however. What is it?	1
	That means that 14 and 15 are relatively prime.	
	We have used three ways to find the greatest common factor of two numbers. What are they?	1. Make a list of all the factors of each. Find the largest one on both lists. 2. Make a list of all the factors of the smaller number. Test each of them on the other. 3. Write down the prime factorizations of each number. Find the largest string that is in both.

TEACHER ACTION	TEACHER TALK	EXPECTED RESPONSE
Students need Worksheet 6-2, Greatest Common Factors.	On Worksheet 6-2 use prime factorization to find the GCF of the given numbers. You are also given a story problem situation and asked to answer two different questions about it.	
Worksheet 6-3, Practice Problems, is provided for use as homework or as an extra challenge sheet.		
An interesting class challenge to use after Activities 5 and 6 or as a review later on is to play the following riddle game.		
Have students draw a number out of a hat. They must write a riddle using the ideas of the unit (plus other previously learned mathematical ideas) that will specify their number completely. Then the class tries to guess students' numbers from their riddles. Some concepts from this unit are factors, factor pairs, products, common multiples, least common multiples, greatest common factors, common factors, primes, relatively prime, divisible, even, odd, and so on.		

NAME

Common Factors

List the factors of each number. Then find the common factors of the two numbers. Do all your work in the space provided.

1. 9 and 24

2. 32 and 48

3. 51 and 17

4. 52 and 8

5. 1,001 and 70

6. 56 and 35

7. The scout leader has a certain number of cookies. They can be divided evenly among 9 scouts. They can also be divided evenly among 6 scouts. What are *two* possibilities for the number of cookies?

8. A band of pirates divided 185 pieces of silver and 148 gold coins. These pirates were known to be absolutely fair about sharing equally. How many pirates were there?

NAME ______________________

Greatest Common Factor

Solve problems 1–4 by finding the prime factorization of each number. Circle the largest string that is in both numbers. Then write the value of the greatest common factor in the box.

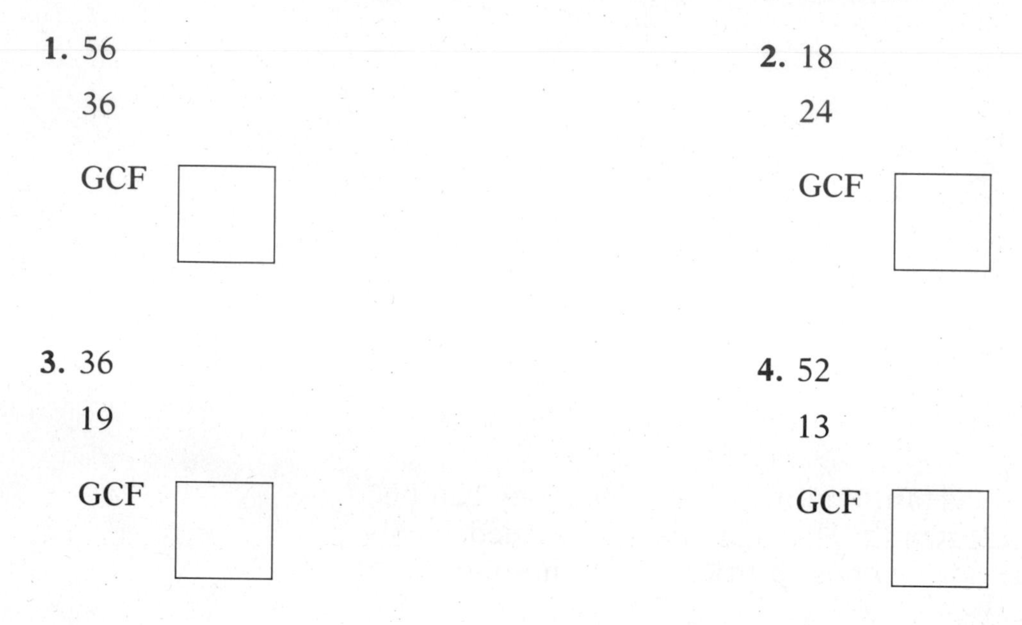

1. 56

36

GCF

2. 18

24

GCF

3. 36

19

GCF

4. 52

13

GCF

Use the following story to solve problems 5 and 6.

Ms. Wurst and Mr. Pop have donated a total of 91 hotdogs and 126 small cans of fruit juice for a math class picnic. Each student will receive the same amount of refreshments.

5. What is the greatest number of students that can attend the picnic?

How many cans of juice will each student receive?

How many hotdogs will each student receive?

6. If one of the hotdogs is eaten by Ms. Wurst's dog just before the picnic, what is the greatest number of students that can attend?

How many hotdogs will each student receive?

How many cans of juice will each student receive?

NAME ______________________

Practice Problems

List all the factors that the following sets of numbers have in common.

1. 21 and 49 ______________________

2. 17 and 37 ______________________

3. 12, 36, and 48 ______________________

4. 92 and 180 ______________________

What is the greatest common factor (GCF) of each of the following numbers?

5. 18 and 36 ______________________

6. 29 and 49 ______________________

7. 165 and 198 ______________________

8. 630 and 1,350 ______________________

9. What is the smallest number that 8 and 12 both divide?

10. a) If 8 and 20 both divide a number N, name four other numbers that must divide N. ______________________

b) What is the smallest number that could be N? ______________________

Activity 7

SIFTING FOR PRIMES

OVERVIEW

Primes were introduced in the Factor Game and have been used in all the activities. This activity explores the classical "Sieve of Eratosthenes" for obtaining prime numbers. When created with colored pens, however, the sieve also contains much information about factors, multiples, and composite numbers; so extensive review of previous concepts is provided through the follow-up questions for the sieve. A simple application of primes is used when students are asked to decipher a code in Worksheet 7-4.

Goals for students

1. Review the definition of prime and composite numbers.
2. Review the relationship between factor and multiple: if $5 \times 6 = 30$, 5 and 6 are factors of 30, and 30 is a multiple of 5 and 6.
3. Use the technique of sifting to find primes.
4. Use square numbers to predict the last number that must be sifted to find all primes less than a given number.

Materials

Calculators.

Colored transparency pens.

Crayons or colored pens (optional).

Worksheets

*7-1, 100 Board.

*7-2, 101-200 Board.

7-3, Using the Sieve.

7-4, Prime Puzzle.

7-5, Practice Problems.

Transparencies

Starred items should be made into transparencies.

SIFTING FOR PRIMES

TEACHER ACTION	TEACHER TALK	EXPECTED RESPONSE
Students need calculators; Worksheet 7-1, 100 Board; crayons or colored pens.	When we played the factor game there were several first moves that scored only 1 point for the opponent.	
	What were some of these?	29, 33, 19, 17, and so on.
	What did we call these?	Prime numbers.
	Does anyone know what a sieve is?	A strainer made of wire mesh; a colander for draining vegetables or pasta.
	What would happen if you sifted sand at the beach through a sieve to look for something you lost?	The sand would fall through the holes and leave your treasure in the sieve.
	About 2,000 years ago a Greek mathematician named Eratosthenes invented a way to sift to find prime numbers. Today we are going to use Eratosthenes's technique. We will sift numbers.	
	At the end of our sifting, the primes will be left.	
You may want to have students review multiples if necessary, using a calculator.	First we need to do some exercises on our calculators.	
Most inexpensive calculators have an automatic constant for addition. This allows us to use 0 + 2 [=] [=] [=] and so on to generate multiples. If your calculators do not have this feature, then substitute the key sequence 2 + 2 [=] + 2 [=] + 2 [=] + 2 [=] and so on to generate multiples.	Press 0 + 2 [=] [=] [=] and so on.	
	What do you see?	2, 4, 6, 8, and so on. The even numbers.
		The multiples of 2. Count by two's.
	Let's write these numbers showing that 2 is a factor of each.	Numbers that have 2 as a factor.
		$2 \times 1 = 2, 2 \times 2 = 4, 2 \times 3 = 6, 2 \times 4 = 8$, and so on.

TEACHER ACTION	TEACHER TALK	EXPECTED RESPONSE
	Clear your calculator. What will happen if we press 0 + 5 [=] [=] [=] and so on?	We'll get 5, 10, 15, 20, and so on.
	Try it.	
	What did you get? Tell me how they are multiples.	$5 \times 1 = 5, 5 \times 2 = 10, 5 \times 3 = 15$, and so on.
	What would you do to get 3, 6, 9, 12, 15, and so on?	Press 0 + 3 [=] [=] [=] and so on.
Display a transparency of Worksheet 7-1, 100 Board, on the overhead. Give the class instructions, in appropriate colors, on the overhead. 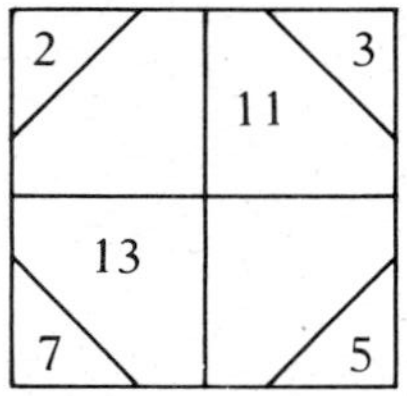	Cross out the 1. One is not a prime.	
2, 3, 11, 13, 7, 5	The next number is 2. It is a prime. Circle it in red. What are the multiples of 2? Shade the upper left corner of the other multiples of 2 in red. Do not shade 2 itself.	2, 4, 6, 8, and so on. 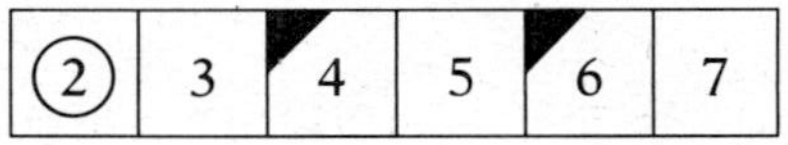 (2) 3 4 5 6 7 and so on.
	Where are the multiples of 2?	They're in columns.
	We'll say that the shaded numbers fell through our sieve. What numbers are they?	The even numbers (except for 2).
The overall scheme for marking the 100-Board is illustrated in this square. With older students the marks alone will suffice without coloring in the triangular areas. Whichever scheme you use, have students record in the square provided on Worksheet 7-1 the code used to make the sieve. The follow-up questions make it *necessary* for students to be able to read the code on a particular number after the sieve is finished.	The next number is 3. What are the multiples of 3? Circle 3 in green.	3, 6, 9, 12, and so on.
	Starting with 6, shade the other multiples of 3 using your green pen; shade the upper right corner.	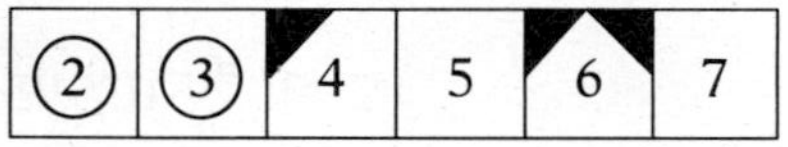 (2) (3) 4 5 6 7 and so on.
	Where are the multiples of 3?	On diagonals.
	The next number is 4. It's already been shaded. Why?	It is a multiple of 2.

TEACHER ACTION	TEACHER TALK	EXPECTED RESPONSE
	What is the next number that is clear? Circle it in orange.	5.
	Sift out the multiples of 5 by shading in orange the bottom right corner of the square for each multiple of 5.	10 20 and so on.
	What was the first multiple of 5 that wasn't already sifted out?	25.
	25 = 5 × 5. When did 5 × 2 get sifted out?	In sifting for 2s.
	When did 5 × 3 get sifted out?	In sifting for 3s.
	When did 5 × 4 get sifted out? Look at the colors.	In sifting for 2s.
Do the same for 7. Use blue in the lower left corner.	Circle 7 in blue. What are the multiples of 7 that aren't already sifted out?	49 = 7 × 7 77 = 7 × 11 91 = 7 × 13
	When did 7 × 10 fall through?	When we sifted 2s and again when we sifted 5s.
	When did 7 × 12 fall through?	When we sifted 2s and 3s.
	What's the next clear number?	11
	What would be the first multiple of 11 to fall through on the sifting?	11 × 11 = 121
	Is it on our board?	No.
	11 × 2 fell through when we sifted for 2s.	
	11 × 3 fell through when we sifted for 3s.	
	What about 11 × 6?	It fell through with 2s and 3s.
	What about 11 × 10?	It's too big. It would have fallen through on 2 and on 5.

TEACHER ACTION	TEACHER TALK	EXPECTED RESPONSE
	11×11 is beyond our chart.	
	Suppose we do sift for multiples of 11.	
	What is the next uncircled number after 11?	13.
	If we sifted for multiples of 13, what would be the first number to fall through?	$13 \times 13 = 169$
	This is also beyond our chart. If we continue to sift higher numbers, will any of the remaining numbers in our chart be eliminated?	No!
	When could we stop sifting on the 100 board? Why?	The first multiple of a number not already sifted is the square of the number. $7 \times 7 = 49$, which is on the board, but $11 \times 11 = 121$, which is beyond the board. So we can stop at 7.
List all the primes, in order, at the bottom of the sheet.	Circle all the remaining numbers in black. All the circled numbers are primes	2, 3, 5, 7, 11, 13, 17, 19, 23, 29, 31, 37, 41, 43, 47, 53, 59, 61, 67, 71, 73, 79, 83, 89, 97.
	Look at all the primes we got just by sifting as far as 7.	
	How many primes from 1 to 50?	15.
	How many from 50 to 100?	10.
Record some of the guesses.	How many would you expect from 100 to 150?	Various answers.
	From 150 to 200?	Various answers.
	How would you find out for sure?	Keep sifting.

<table>
<tr><th>TEACHER ACTION</th><th>TEACHER TALK</th><th>EXPECTED RESPONSE</th></tr>
<tr><td>Students need Worksheet 7-2, 101–200 Board. Fifth and sixth graders will need to be paced through this sieve by the teacher. Seventh and eighth graders can be assigned the task to do individually.</td><td>On Worksheet 7-2 are the numbers 101 to 200. We will go back to multiples of 2, then 3, then 5 and so on to extend your sieve to 200.</td><td></td></tr>
<tr><td></td><td>How far will we have to sift?</td><td>Various answers.</td></tr>
<tr><td>Use a vertical line to mark multiples of 11 and a horizontal for multiples of 13.
<table><tr><td></td><td>11</td></tr><tr><td>13</td><td></td></tr></table></td><td>We saw on the 100 board that the first multiple of a number that wasn't already sifted out was the square of the number. We sifted for 7s because $7 \times 7 = 49$, $7 \times 11 = 77$, and $7 \times 13 = 91$ were not sifted out. But $11 \times 11 = 121$ was beyond 100.</td><td></td></tr>
<tr><td></td><td>We will have to sift for 11s on the 200 chart. What about 13s?</td><td>Yes; $13 \times 13 = 169$.</td></tr>
<tr><td></td><td>What is the next prime after 13?</td><td>17.</td></tr>
<tr><td></td><td>Will we need to sift for 17s?</td><td>No. $17 \times 17 = 289$.</td></tr>
<tr><td></td><td>So we must sift for 2, 3, 5, 7, 11, and 13 on our new board to find all the primes. What are the new primes?</td><td>101, 103, 107, 109, 113, 127, 131, 137, 139, 149—10 primes between 100 and 150; 151, 157, 163, 167, 173, 179, 181, 191, 193, 197, 199—11 primes between 150 and 200.</td></tr>
<tr><td>You may have to keep reminding students by asking more questions about why we know we can stop with 17. 17×2, 17×3 up to 17×16 have already been sifted out. The next prime is 19; all the multiples 2×19 up to 17×19 have been sifted out, and 19×19 is not on the board.</td><td>How far would we have to sift on a 300 Board? We would only have to sift for one more prime if we extend the board to 300.</td><td>17 because $17 \times 17 = 289$; 19 is too large, $19 \times 19 = 361$.</td></tr>
</table>

TEACHER ACTION	TEACHER TALK	EXPECTED RESPONSE
As an extra challenge for more advanced students who will have to devise a strategy, ask.	What's the largest prime less than 1,000?	997.
	What's the smallest prime greater than 1,000?	1,009.
With older students, you may want to use the square root button on the calculator to find how far to sift to get primes through 200. We would sift through the largest prime less than or equal to the square root of 200. Then you could predict how far to go for larger boards. For example: How far do we sift to get primes ≤ 500? (through 19) How far do we sift to get primes ≤ 1,000? (through 31)		

TEACHER ACTION	TEACHER TALK	EXPECTED RESPONSE
Record the number of primes from 100 to 150 (10) and from 150 to 200 (11): 101, 103, 107, 109, 113, 127, 131, 137, 139, 149; and 151, 157, 163, 167, 173, 179, 181, 191, 193, 197, 199.	We often look for patterns in mathematics. Do these numbers appear to follow any pattern?	No.
	Look at the row starting with 181.	Only one prime. There are 9 composites in a row.
Ask.	Mathematicians have not discovered a pattern that includes exactly the prime numbers.	
	How far did you have to go to find all the primes less than 200? Why?	Through 13. $13 \times 13 = 169$ $15 \times 15 = 225$
	How far would you have to go to find all the primes less than 1,000? Why?	Through 31. $31 \times 31 = 961$ $37 \times 37 = 1{,}369$ The square of 31 is on the board. The square of 37 is not.
	Our chart does more than find primes.	
	Find some numbers that are marked with the code for both 2 and for 5.	20, 10, 50, and so on.
	How did you tell?	Looked for numbers with red triangle in upper left corner *and* orange triangle in lower right corner.
	These numbers are multiples of 2 and of 5 at the same time. What numbers are multiples of both 3 and 7?	21, 42, 63, 84, and so on.
	What numbers are multiples of both 5 and 6?	30, 60
	How did you tell by looking at colors?	Multiples of 2, 3, and 5 have red, green, and orange shading.

TEACHER ACTION	TEACHER TALK	EXPECTED RESPONSE
	What primes are factors of 72? Why?	2, 3; $2 \times 36 = 72$; $3 \times 24 = 72$. 72 is shaded only with the code for 2 and 3.
Worksheet 7-3, Using the Sieve, and Worksheet 7-5, Practice Problems, are to be used as homework. Worksheet 7-4, Prime Puzzle, is enjoyed by all students.		
Refer to the 100 and 101–200 Boards.	How can we find the prime factorization of 72?	2 and 3 are the only prime factors and by dividing we find that $72 = 2 \times 2 \times 2 \times 3 \times 3$.
	Do 72 and 54 have any prime factors in common?	Yes. 2 and 3.
	What is the GCF for 72 and 54?	$2 \times 3 \times 3$, or 18
	Which numbers are multiples of 3 distinct primes?	3 primes: 30, 42, 60, 70, 84, 90, 105, 110, 126, 130, 132, 140, 154, 165, 168, 176, 180, 182, 195, 198.
	Which are multiples of 4 distinct primes?	4 primes: none (note: $2 \times 3 \times 5 \times 7 > 200$)
	Which are multiples of 5 distinct primes?	5 primes: none

NAME ______________________

100 Board

Code

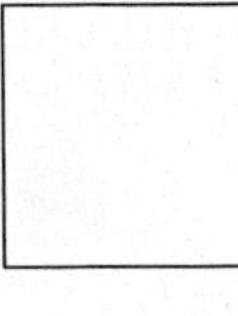

1	2	3	4	5	6	7	8	9	10
11	12	13	14	15	16	17	18	19	20
21	22	23	24	25	26	27	28	29	30
31	32	33	34	35	36	37	38	39	40
41	42	43	44	45	46	47	48	49	50
51	52	53	54	55	56	57	58	59	60
61	62	63	64	65	66	67	68	69	70
71	72	73	74	75	76	77	78	79	80
81	82	83	84	85	86	87	88	89	90
91	92	93	94	95	96	97	98	99	100

NAME

101–200 Board

Code

101	102	103	104	105	106	107	108	109	110
111	112	113	114	115	116	117	118	119	120
121	122	123	124	125	126	127	128	129	130
131	132	133	134	135	136	137	138	139	140
141	142	143	144	145	146	147	148	149	150
151	152	153	154	155	156	157	158	159	160
161	162	163	164	165	166	167	168	169	170
171	172	173	174	175	176	177	178	179	180
181	182	183	184	185	186	187	188	189	190
191	192	193	194	195	196	197	198	199	200

NAME

Using the Sieve

Use your 100 Board to answer the following questions.

1. What is the smallest prime number that is greater than 30?

2. What is the smallest prime number that is greater than 50?

3. 5 and 7 are called *twin primes* because they are both primes and they differ by two. List all twin primes between 1 and 100.

4. A number that is not prime (and not 1) is a composite number. Find 5 composite numbers in a row.

5. Why didn't we sift for 9s?

6. Which of the primes 2, 3, 5, and 7 divide 84?

7. The number 6 was sifted with both 2 and 3.

a) Find all other numbers that were sifted with both 2 and 3.

b) How are all the multiples of 6 marked on the board?

8. **a)** List the multiples of 7.

b) What numbers are multiples of *both* 6 and 7.

9. How are multiples of 15 marked on the board?

10. There are four columns on the board that contain no primes. Find them and explain why these columns contain no primes.

NAME ______________________

Prime Puzzle

There is a message hidden below. Cross out the letters in the boxes containing numbers that are *not* prime numbers to discover the message in the remaining boxes.

D 7	P 6	I 2	R 8	V 19	I 11	M 12	P 60	S 3	K 9	S 14	O 59	Z 35	R 11	S 37
Q 4	A 3	R 31	M 25	E 23	S 10	D 29	M 12	I 41	V 97	H 100	I 23	N 83	E 13	A 12
B 71	U 2	R 35	T 3	T 27	F 43	O 42	A 37	I 64	C 7	T 5	R 45	O 13	R 11	S 71
N 9	E 14	U 69	M 32	A 17	S 87	F 48	G 75	O 20	R 19	K 9	E 97	Q 8	T 27	D 57
F 67	R 2	C 16	I 89	M 18	E 7	T 12	K 9	N 17	D 73	L 67	N 49	I 59	E 29	R 83

NAME ____________________

Practice Problems

Use your 100–200 Board to answer the following questions.

1. List all the primes between 100 and 200.

2. What prime numbers are factors of 84?

3. What prime numbers are factors of 110?

4. List all the multiples of 7 that are between 130 and 200.

5. Why didn't we sift for 17s on the 200 board?

6. A number that is not prime (and not 1) is a composite number. Find 5 composite numbers in a row on the 200 board. What is the longest string you can find?

7. Why is it impossible to find a number on the 200 board that is divisible by four different prime numbers? (Hint: How big would such a number have to be?)

8. What would be the last prime we would have to sift to find all primes less than 1,000?

Activity 8

PAPER POOL

OVERVIEW

Paper Pool is an application involving many of the concepts encountered in the unit: factors, multiples, rectangles, the relation of being relatively prime. Before seeing how to apply these concepts, students must gather and organize data, then search for patterns.

Paper Pool is played with an imaginary ball being hit from the lower left-hand corner marked A, at a 45° angle. A ball hit in this way will bounce off the sides at a 45° angle. Also, if a grid is placed on the table, the ball always traverses on diagonals of the squares of the grid. For example, the illustration below shows the path of a ball on a 4×6 table. The ball ends up in pocket D; there are five hits and 12 squares are crossed.

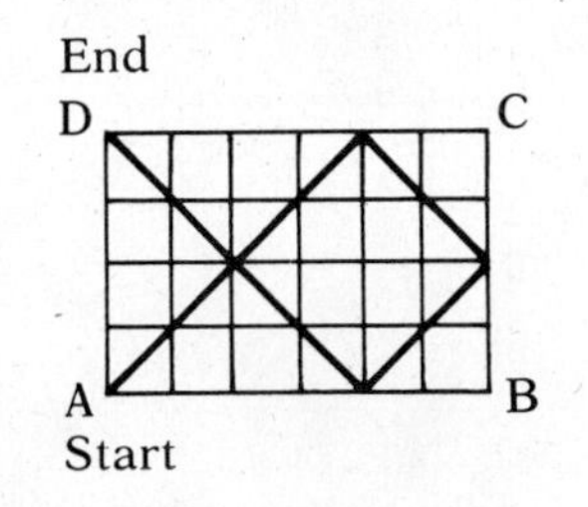

Students first learn how to predict the pocket into which a ball will fall and the number of hits as the ball crosses the table. Students further develop their analytical skills by investigating the number of squares crossed. All three relations depend upon the lengths of the sides of the pool table.

Goals for students

1. Recognize rectangles whose sides have the same ratio.
2. Use the concept of common factor to find the rectangle with the smallest area having a given ratio of sides.
3. Practice organizing data and looking for patterns.
4. Use the concept of common factor and a parity check to predict the behavior of a ball on a pool table: final corner, number of hits.

Materials

Colored pencils
*Basic Table (Materials 8-1).
*Overlay 1 (Materials 8-2).
*Overlay 2 (Materials 8-3).

Worksheets

*8-1, Introduction to Paper Pool.
8-2, Paper Pool, pages 1, 2, and 3.
*8-3, Recording Sheet.
8-4, Supertables.
8-5, Advanced Paper Pool.
8-6, How many Squares Are Crossed?

Transparencies

Starred items should be made into transparencies.

PAPER POOL

TEACHER ACTION	TEACHER TALK	EXPECTED RESPONSE
Students need Worksheet 8-1, Introduction to Paper Pool.	Today we are going to learn to play paper pool. Our pool tables will be grid paper rectangles marked off in squares. There is a pocket at each corner of the tables.	
Display a transparency of Worksheet 8-1, Introduction to Paper Pool, on the overhead. Cover up the bottom 10 × 4 table. 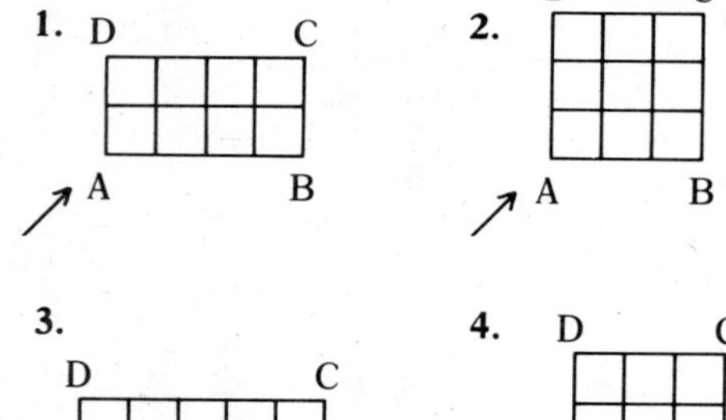	Here are several pool tables. We will label the corners A, B, C, D and always in the order shown. When telling the size of a table we will always give the bottom edge (number of squares across) then the side edge (vertical number of squares).	
Enter bottom edge and side edge for Table 1. Put side edge along right hand edge. 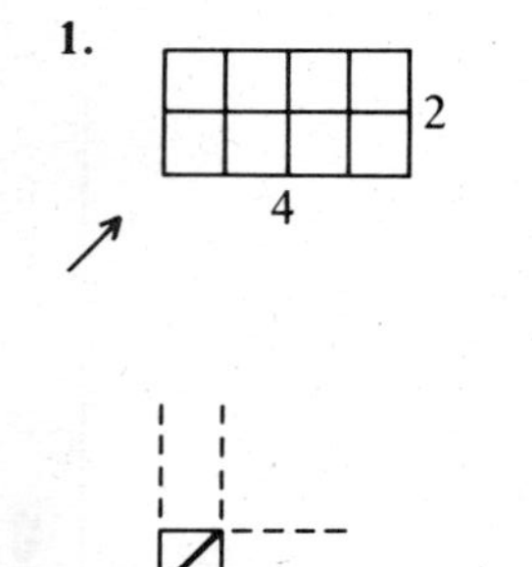	What is the bottom edge of Table 1?	4
	What is the side edge of Table 1?	2
	We'll write 4 × 2, "four by two," for the dimensions of Table 1.	
	What are the dimensions of the other tables?	Table 2 is 3 × 3. Table 3 is 5 × 1. Table 4 is 3 × 2.
	We are always going to start the ball on our pool tables from corner A, the lower left-hand corner. To start the ball we hit it with an imaginary cue-stick. The ball will travel diagonally across the grid square. It continues in a straight line until it hits an edge. This would be like hitting the side cushion of a pool table.	

TEACHER ACTION	TEACHER TALK	EXPECTED RESPONSE
Draw a path across the first two squares of Table 1.	What will happen to the ball?	The ball will bounce off the edge.
	What will be the path of the ball after it hits the edge?	Various answers. Some students may know that the ball will bounce off the edge at the same angle. Illustrate what will happen emphasizing that the path always cuts the small squares on a diagonal.
Draw the path slowly, one square at a time, emphasizing the diagonals that make up the path.	We've hit a pocket so the ball stops. Which corner are we at?	Corner B.
	There have been several hits: the cue hit the ball, the ball hit the edge, the ball hit the pocket.	
	How many hits in all?	Three.
Give the students a minute to draw the path and then illustrate on the overhead, emphasizing number of hits and the pocket where the path ends.	Now try drawing the path of the ball on Table 2.	
Draw paths, count total hits, and mark the corner into which the ball falls for Table 2.		
Table 2. 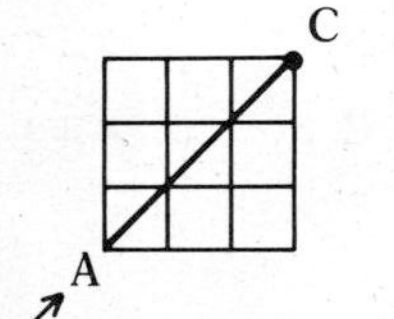	How many hits?	Two.

TEACHER ACTION	TEACHER TALK	EXPECTED RESPONSE
Repeat for tables 3 and 4.		
Table 3. 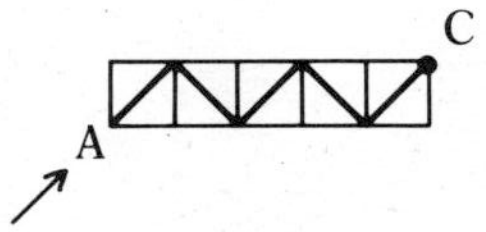 (A, C)	How many hits?	Six.
Table 4. 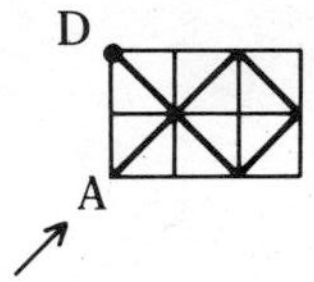 (D, A)	How many hits?	Five.
Now reveal the 10 × 4 table on the overhead. Call on several students to predict the final corner and number of hits. Record the predictions. 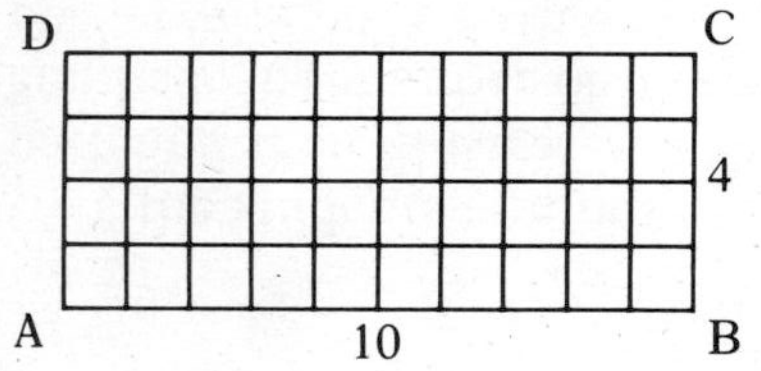 (D, C, A, B; 10, 4)	What size is this table?	10 by 4.
	Before we draw the path on this big one, let's predict where the ball will fall *and* how many hits. Who has a prediction?	Various predictions.
	Now I'll make a prediction: (1) The ball will end in corner D. (2) There will be seven hits.	
Draw and count hits. 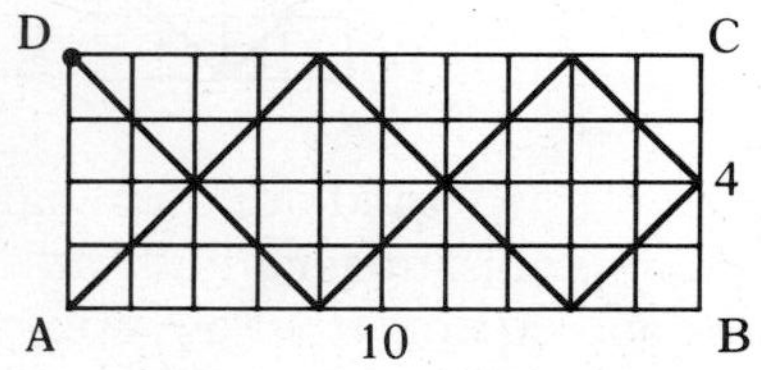 (D, C, A, B; 10, 4)	My predictions were correct!	
	I've got a secret rule that lets me predict what is going to happen on *any* grid table. Your job is to discover my rule.	
	You will need to be systematic. Here are three sheets with tables drawn on them. For each table, very carefully draw the path of the ball, and record both the final corner and the total number of hits.	
Students need Worksheet 8-2, Paper Pool, pages 1, 2, and 3.	What do we count as hits?	When the ball hits an edge. When the cue hits the ball. When the ball hits the pocket.
	Use a colored pencil so that the paths show up.	

OBSERVATIONS	POSSIBLE RESPONSES
Give Worksheet 8-4, Supertables, as an extra challenge. S1: 22 × 18; corner C; 20 hits S2: 24 × 14; corner B; 19 hits	
As an extra challenge, give students grid paper to make additional tables and ask them to find a table with exactly 15 hits. Many answers are possible: 14 × 1, 13 × 2, 11 × 4, 8 × 7.	As the students are working, check to see that they are drawing paths correctly.

TEACHER ACTION	TEACHER TALK	EXPECTED RESPONSE
Students need Worksheet 8-3, Record Sheet.	In order to see patterns that will help us to develop rules for the number of hits and the corner where the path ends, we need to carefully organize all the data that we have from the tables we have done so far.	
Put a transparency of Worksheet 8-3 on the overhead.	Look at table in Problem 1 on your sheet. For this table there were three hits with the ball landing in pocket D. Remember that we list the bottom edge first and then the side edge. What is the size of Table 1?	1×2.
Record in the appropriate space on the transparency. 1×2 2×4 3×6 4×8 5×10 6×12 8×16 9×18 10×20 etc.	What other tables have three hits and land in pocket D?	2×4; 3×6; 4×8.
	What do you notice about the path of the ball on each of these tables?	They all have the same path pattern.
	What are some other tables that would have three hits and land in pocket D?	Various answers. Accept and record only those of the form $K \times 2K$.
	What is the smallest table in this group?	1×2.
Circle 1×2 in red.	1×2 is the *basic table* for this family of tables.	
	How do we get other members of the family from the basic table?	They are multiples of the basic table.
	If a table had a bottom edge of 32, what would its side edge be if it belonged to this family?	64.
	Is an 18×38 table a member of this family?	No; $2 \times 18 \neq 38$.

TEACHER ACTION	TEACHER TALK	EXPECTED RESPONSE
Ask.	Look at Table 5. What is the size of the table?	6 × 4.
	Where does it belong on the chart?	Corner D—five hits.
	Find another corner D—5 hits.	Table 6.
	What size is Table 6?	3 × 2.
	Does 6 × 4 belong to the same family as 3 × 2?	Yes.
	What is the basic table for this family?	3 × 2.
	How do you know that it is a basic table?	3 and 2 do not have any common factors. They are relatively prime.
	Name some other tables in this family.	Accept multiples of 3 × 2 (6 × 4, 9 × 6, 12 × 8, and so on).
Display a transparency of the Basic Table with overlay 1 and 2 on top (Materials 8-1, 8-2, 8-3).	Here is a table with its path drawn. What size is it?	12 × 18.
	The ball lands in pocket B with five hits.	
Pull back overlay 2 to reveal the 4 × 6 table.	The path is still exactly the same pattern; the ball lands in pocket B with five hits because *the path does not change within a family of tables.*	
	What size is this new table?	4 × 6.
	Is this a basic table?	No, because 4 and 6 have a factor in common.
Pull back overlay 1 to reveal the 2 × 3 table.	What would be the basic table for this family?	2 × 3.

TEACHER ACTION	TEACHER TALK	EXPECTED RESPONSE
List several tables in column B-5. 5 B 2 × 3 4 × 6 12 × 18 6 × 9 and so on.		
Collect remaining data either together as a group, recording on the overhead *and* on the students' copies, or you can have the students complete the recording and quickly check the results.	Let's record all the remaining data from the tables on our sheet. Circle each basic table.	
Ask.	What basic tables are in column C-8?	5 × 3 1 × 7 3 × 5
	Are there other columns with more than one basic table?	Yes. B-7 and B-9.
	So columns can have more than one family. Name another basic table that would go in B-7.	6 × 1. If the students do not see this one, have them draw and try out a 6 × 1 table.
Ask.	Look at the Record Sheet as a whole.	
	Do you see any patterns?	Various answers; students may see that a table with 2, 4, 6, 8, etc., hits (an even number of hits) always ends in corner C; tables with an odd number of hits never end in C; the reverse of tables that end in B end in D: 4 × 2 ends in B, 2 × 4 ends in D.
	Let's find a rule for the number of hits.	
	Name a basic table with 12 hits.	1 × 11, 11 × 1, 7 × 5, 5 × 7.

TEACHER ACTION	TEACHER TALK	EXPECTED RESPONSE
	The dimensions of these tables add up to 12. 2 × 10 3 × 9 4 × 8 5 × 7 6 × 6 Do they have 12 hits?	No. They are not basic tables.
Ask.	What is a basic table with 13 hits?	12 × 1 11 × 2 10 × 3 9 × 4 8 × 5 7 × 6 and reversed
	How could we find the number of hits on a non-basic table?	Find its basic table and add its dimensions.
	How many hits for a 6 × 10 table?	The basic table is 3 × 5, so there are 8 hits.
	Is this what we found by drawing the path?	Yes.
	Now let's look for a rule to predict the final corner.	
	Look at the basic tables that have final corner C. Can you make a statement?	Dimensions are both odd.
	What is the basic table for C-1?	1 × 1.
	What is another basic table for C-10?	9 × 1, 7 × 3
	Let's look at all the basic tables for corner B. What are they?	2 × 3, 4 × 3, 2 × 5, 2 × 7, 4 × 5, and so on.
	What is true of these tables?	The bottom edge is even, the side edge is odd.
	What is true of tables for D?	The bottom edge is odd, the side edge is even.

TEACHER ACTION	TEACHER TALK	EXPECTED RESPONSE
	If I tell you a 10 × 31 table is in B, where is 31 × 10?	31 × 10 is in D; same hits.
	Tell me all about a 30 × 36 table. Is it basic? If not, what is the basic table of the family?	No. 5 × 6.
	How many hits?	11.
	What corner?	D.
	Suppose we have a basic table. How many hits?	The bottom edge plus the side edge.
	What is the final corner?	If both are odd, the corner is C. If the bottom edge is even and the side edge odd, the corner is B. If the bottom edge is odd and the side edge even, the corner is D.
	You've found the secret rule!	
Students need Worksheet 8-5, Advanced Paper Pool. This activity could be used as homework.	Now you get to practice your secret rule on some problems.	
Worksheet 8-6, How Many Squares Are Crossed?, can be used as an extra challenge for younger students.		
Here is another question for students to investigate.	Given the size of the table, how many of the small squares does the ball cross?	It is the LCM of the dimensions.
Check answers to Advanced Paper Pool.		

Basic Table

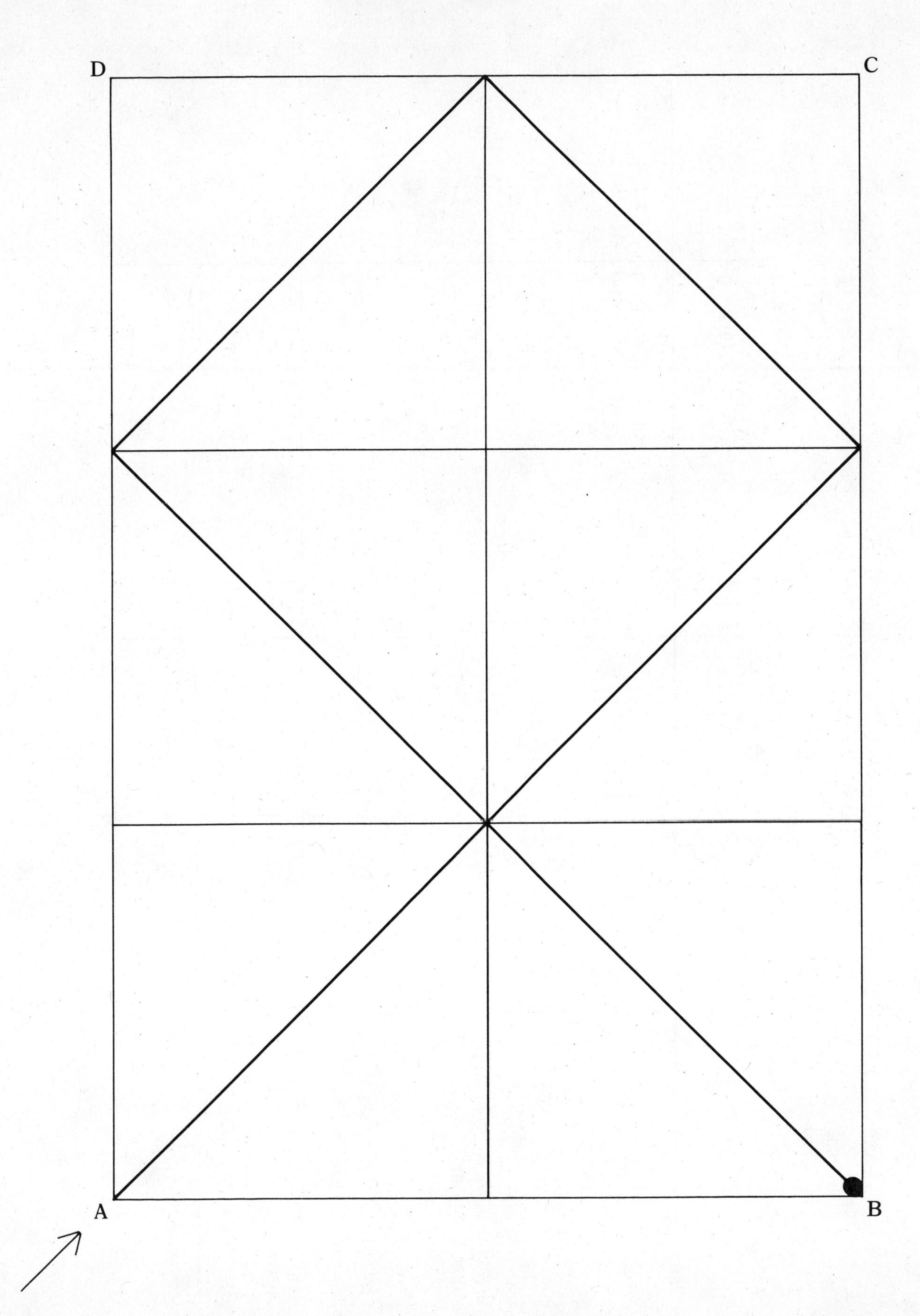

Overlay 1

Overlay 2

NAME

Introduction to Paper Pool

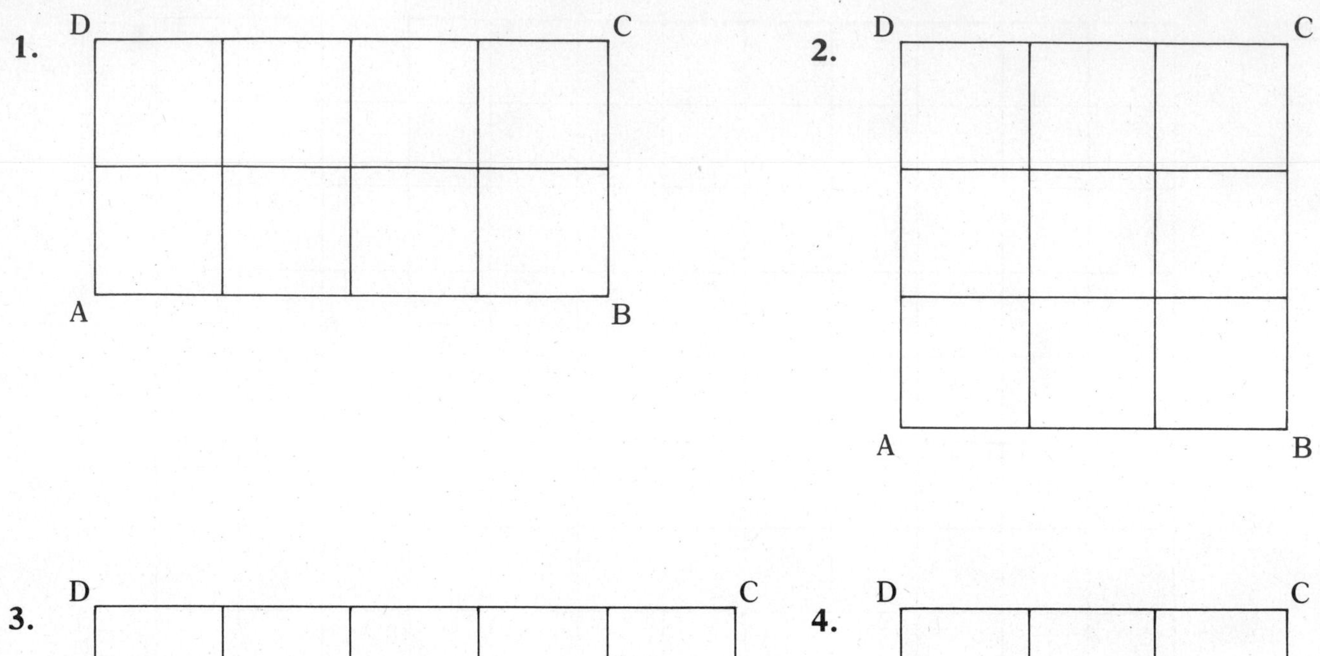

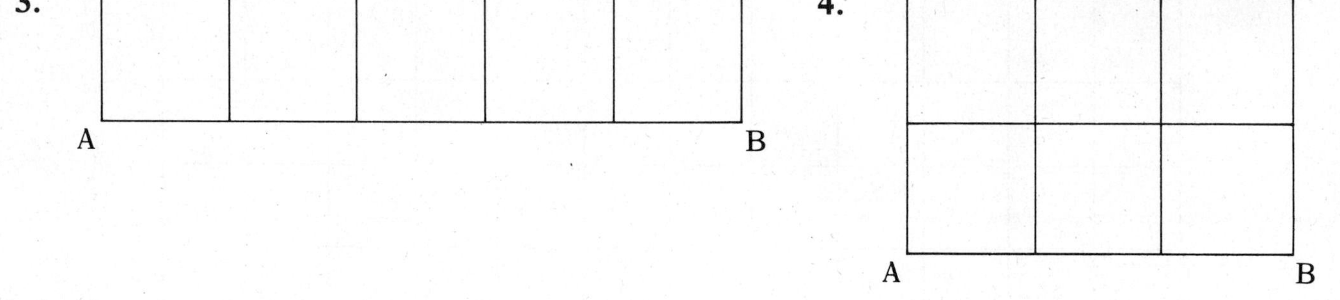

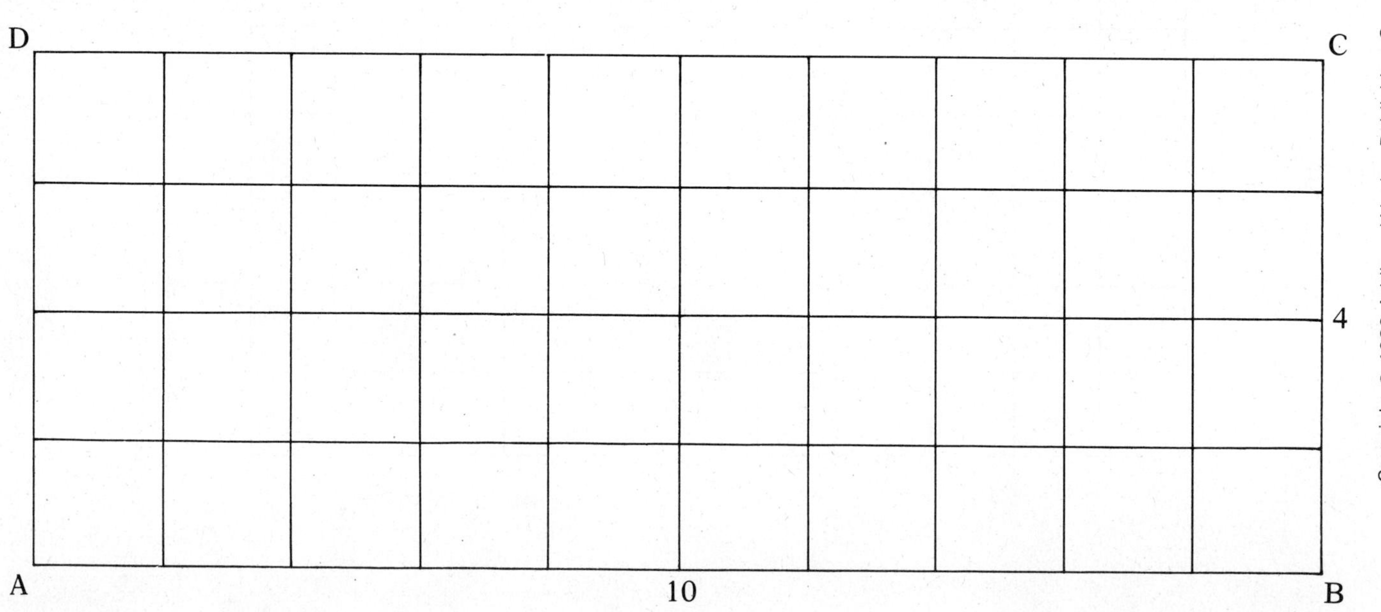

NAME ______________________

Paper Pool

Use a colored pencil to draw the path of the ball.

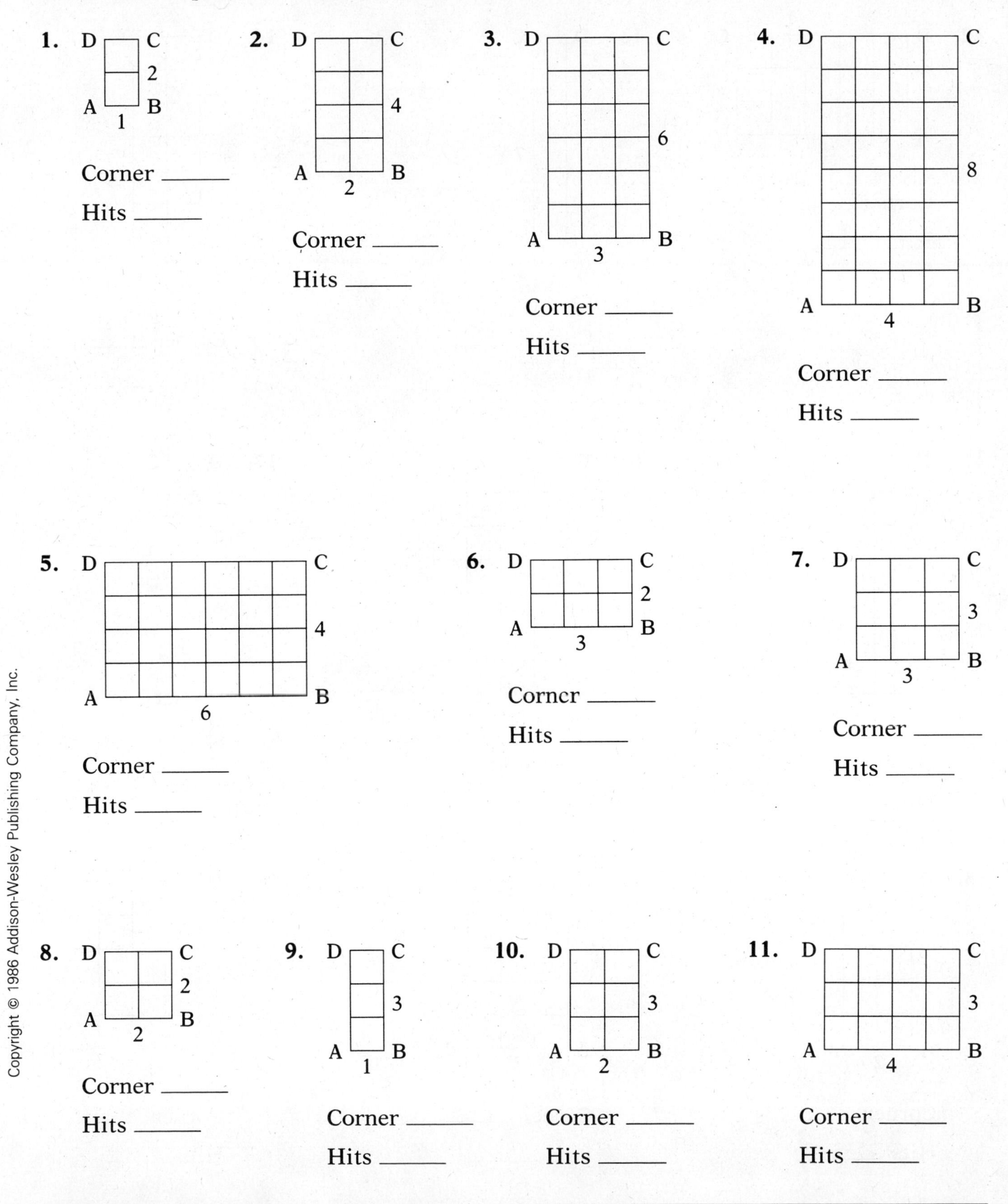

NAME

Paper Pool

Use a colored pencil to draw the path of the ball.

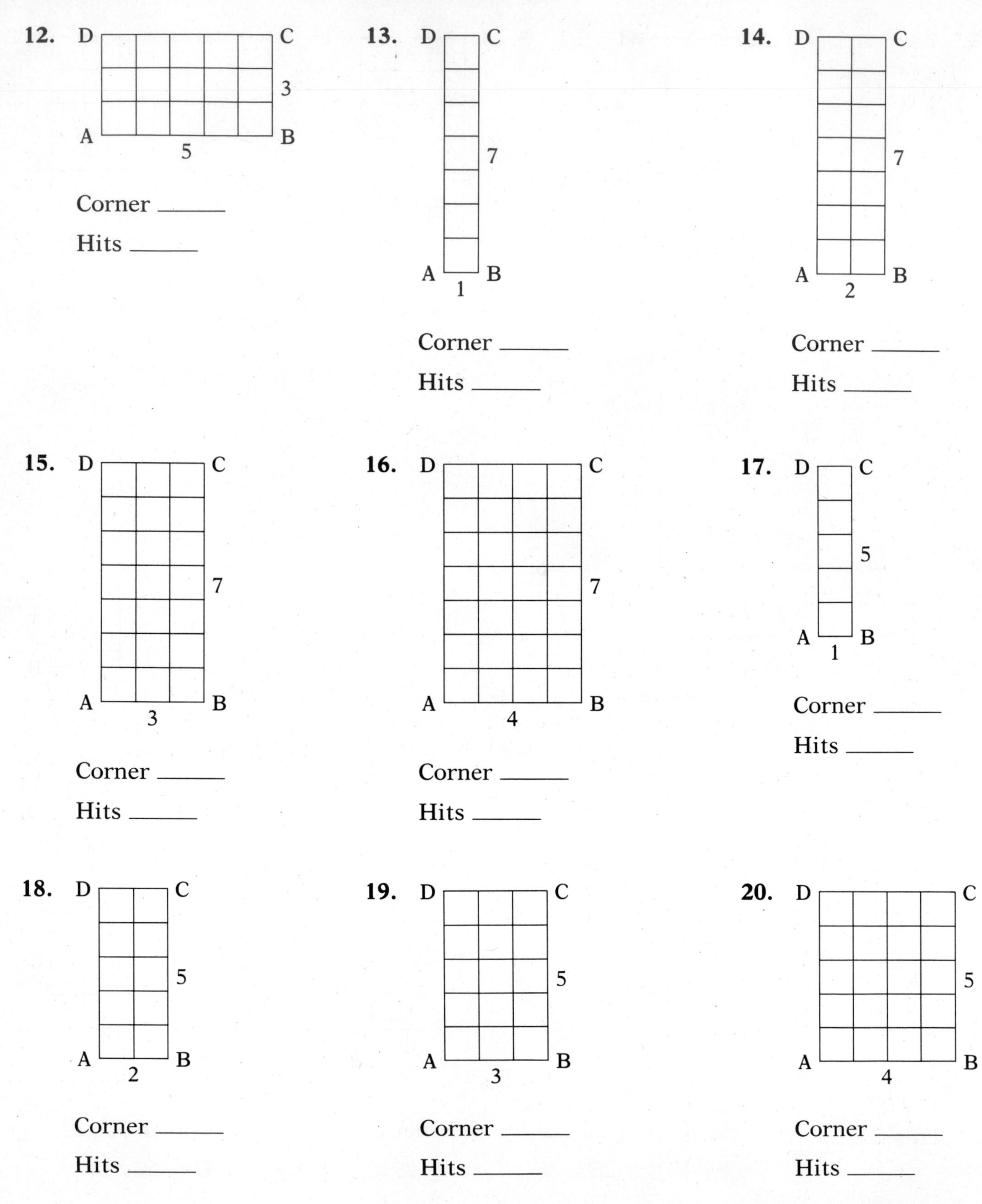

NAME ____________

Paper Pool

Use a colored pencil to draw the path of the ball.

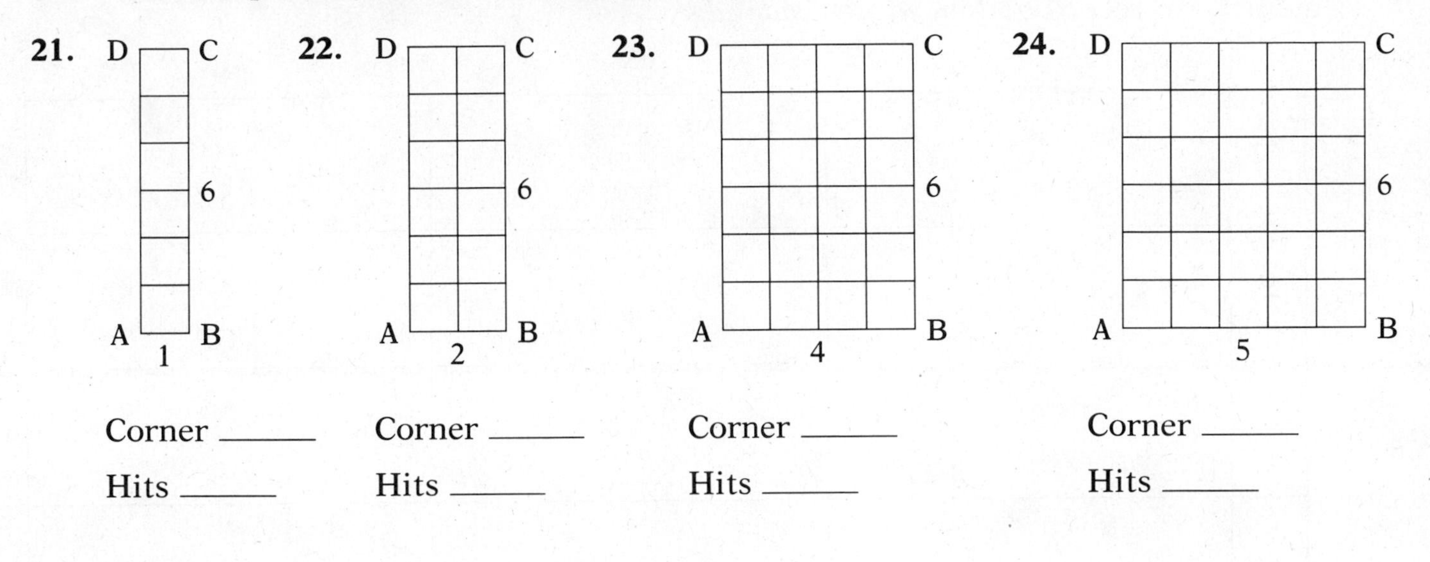

21. Corner ______ Hits ______

22. Corner ______ Hits ______

23. Corner ______ Hits ______

24. Corner ______ Hits ______

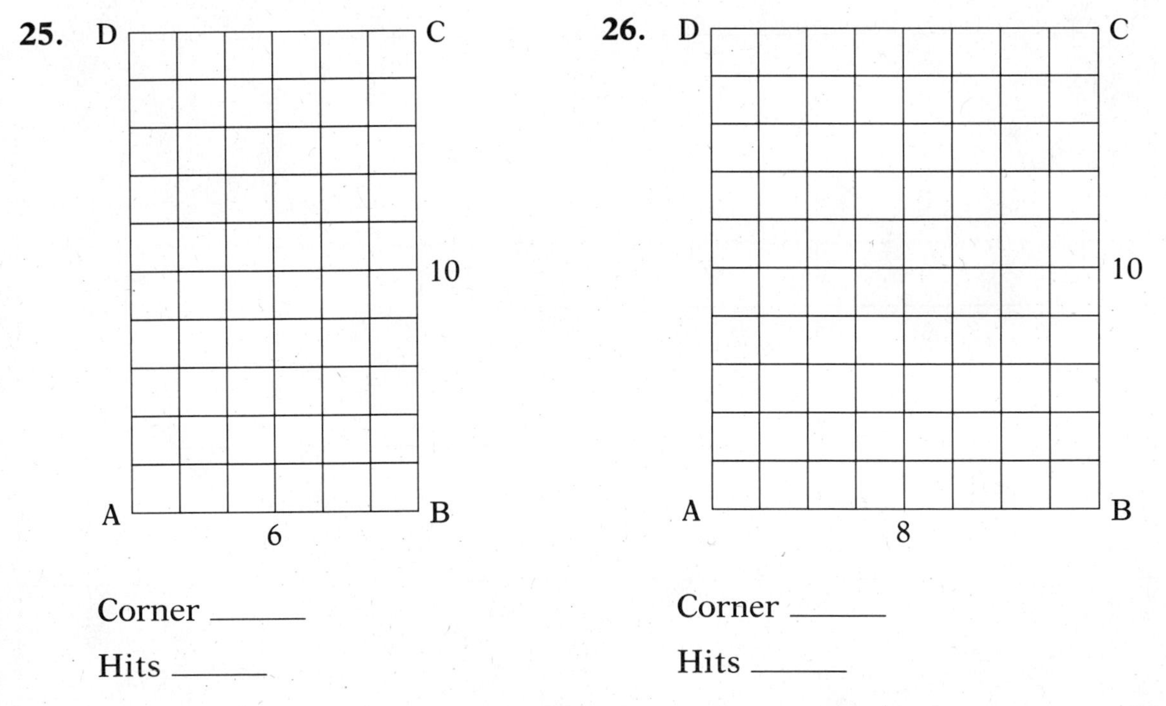

25. Corner ______ Hits ______

26. Corner ______ Hits ______

NAME ____________________

Record Sheet

Record the data from the pool tables in Worksheet 8-2 in this chart.
Remember to record *bottom edge* × *side edge.*

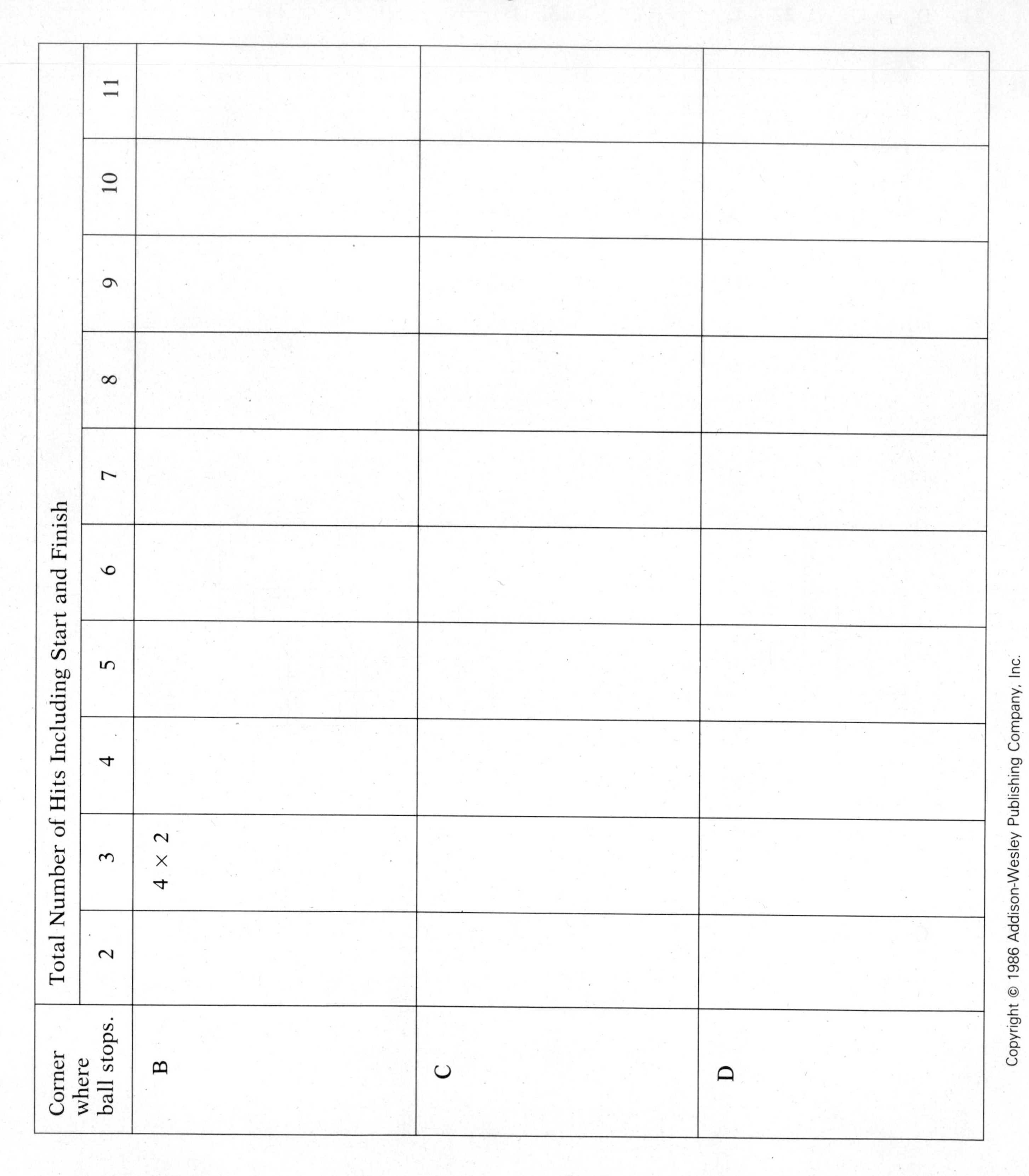

Corner where ball stops.	Total Number of Hits Including Start and Finish									
	2	3	4	5	6	7	8	9	10	11
B		4 × 2								
C										
D										

NAME ______________________

Supertables

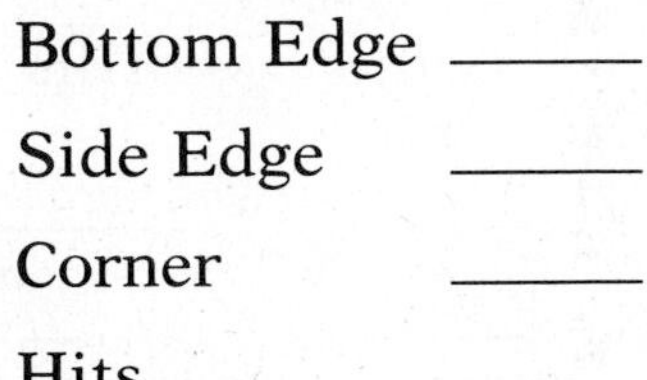

Bottom Edge ______

Side Edge ______

Corner ______

Hits ______

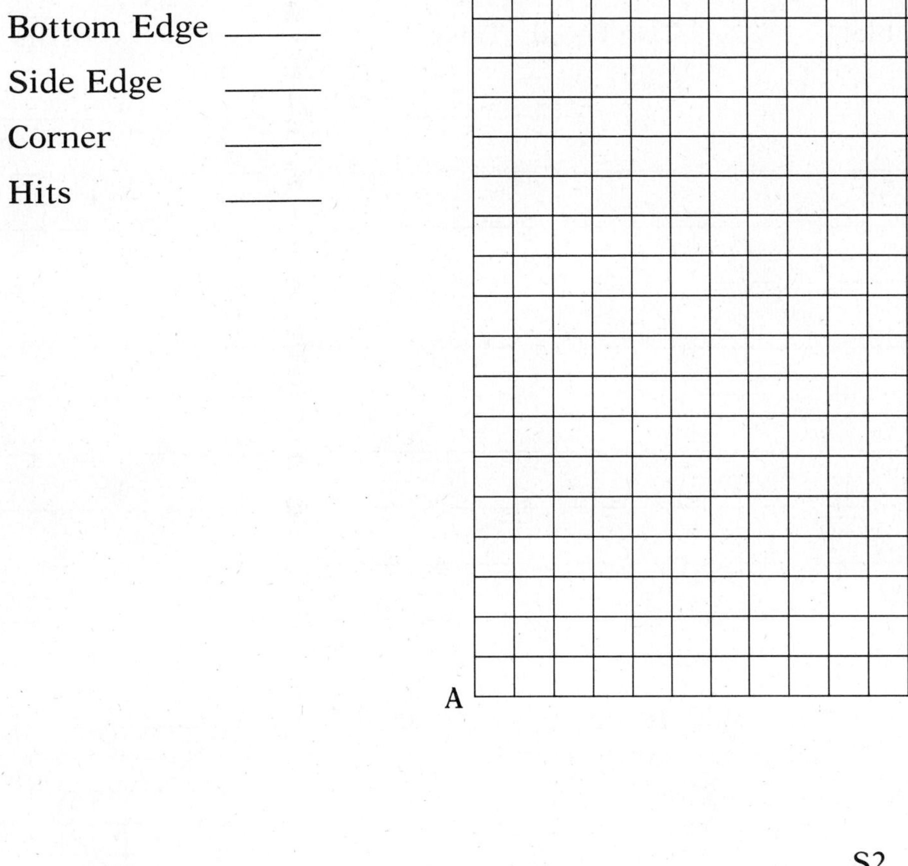

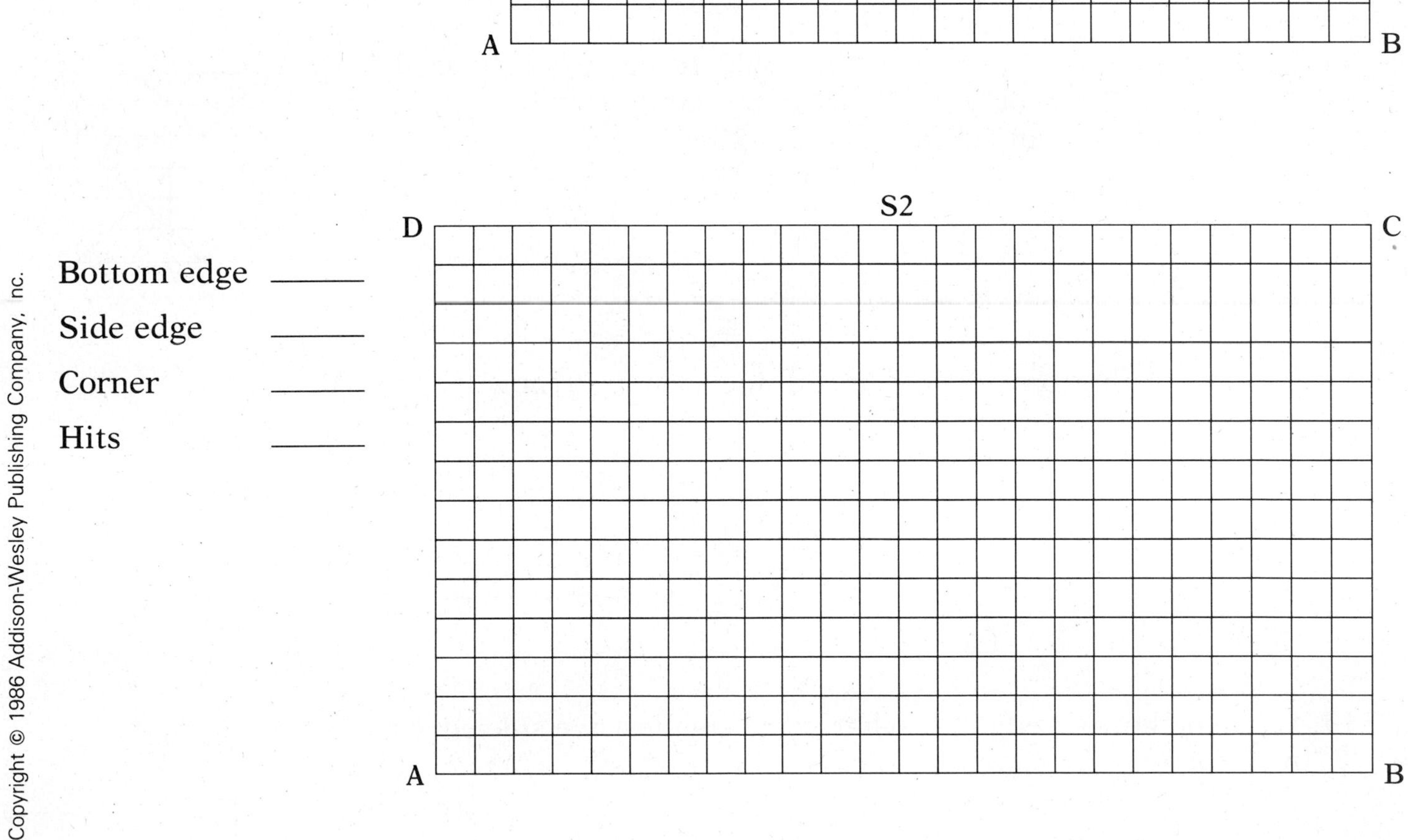

Bottom edge ______

Side edge ______

Corner ______

Hits ______

Advanced Paper Pool

NAME ______________________

For each table, predict the final corner and the number of hits.

	Basic Table	Two Family Tables		Corner	Hits
1. 9 × 11					
2. 33 × 121					
3. 51 × 72					
4. 32 × 72					
5. 45 × 81					
6. 108 × 72					
7. 42 × 7					
8. 13 × 65					
9. 114 × 84					
10. 42 × 13					

11. Draw the path of the ball on this table. If the "6" is changed to "21", in order for the path to be the same, then the "8" should be changed to ______.

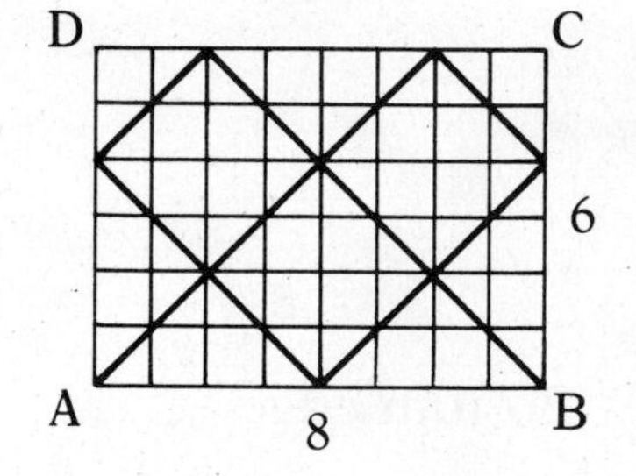

12. A table has bottom edge 14. If the side edge is 40, how many hits are there? ______

D C

A 14 B

13. With the bottom edge 14, what is the smallest the side edge can be in order that there are at least 50 hits? ______

NAME

How Many Squares Are Crossed?

There is also a rule to predict how many squares the path crosses on a paper pool table.

For example: This path crosses two squares. Use your Paper Pool worksheets and record the data from each table.

Size of Table	Number of Squares Crossed

Size of Table	Number of Squares Crossed

Mark the basic tables with a *.

What is the rule for the basic tables?

What is the rule for the other tables?

Does the rule for the other tables also work for basic tables?

How many squares would be crossed in a 24×36 table?

NAME ____________________

Review Problems

1. Find all the factor pairs for the following numbers.

a) 28 **b)** 72

2. What is the smallest number we have to check to be sure that we have all the factor pairs of the following numbers?

a) 60 **b)** 150

3. Decide whether the following numbers are deficient, perfect, or abundant.

a) 12 **b)** 28

c) 21 **d)** 64

4. Fill in the box with a number that will make each statement true.

a) $7 \times \square = 56$

b) $\square \times 3 = 36$

c) $14 \times 5 = \square$

5. Find the factors used to make the following Product Game.

4	■	6	9
10	14	15	■
21	25	35	49

6. Find all the divisors of the following numbers.

a) 30 **b)** 29

NAME ______________________

Review Problems

7. Mrs. Brown wrote a number on the chalkboard and said, "I know that to be sure I find all the factor pairs for this number I must check each number from 1–29." What numbers could Mrs. Brown have written? What is the smallest number Mrs. Brown could have written? What is the largest number Mrs. Brown could have written?

8. A number has exactly five factor pairs. If the number is less than 100 and larger than 50, what is the number?

9. Find the prime factorization for the following numbers.

a) 90 **b)** 71 **c)** 441 **d)** 246

10. a) What is the least common multiple of 6 and 15?

b) What is the greatest common factor of 6 and 15?

11. a) The least common multiple of a number and 12 is 60. What could the number be?

b) The greatest common factor of a number and 12 is 4. What could the number be?

12. a) In checking for all the primes less than 180, how far do you have to sift?

b) List all the primes between 200 and 220.

NAME

Review Problems

13. To find all the primes from 1 to N, you need only sift through 13.

a) What is the smallest number N could be?

b) What is the largest number N could be?

14. In paper pool, a certain table has dimensions 6×15.

a) What pocket will the ball end up in?

b) How many hits are there?

c) How many squares are crossed?

15. If a pool table has a bottom edge of 12, give the dimensions of the side edge so that

a) the ball ends up in pocket B.

b) the ball has at least seven hits.

c) the ball crosses at least 30 squares.

16. Draw all the rectangles with an area of 18.

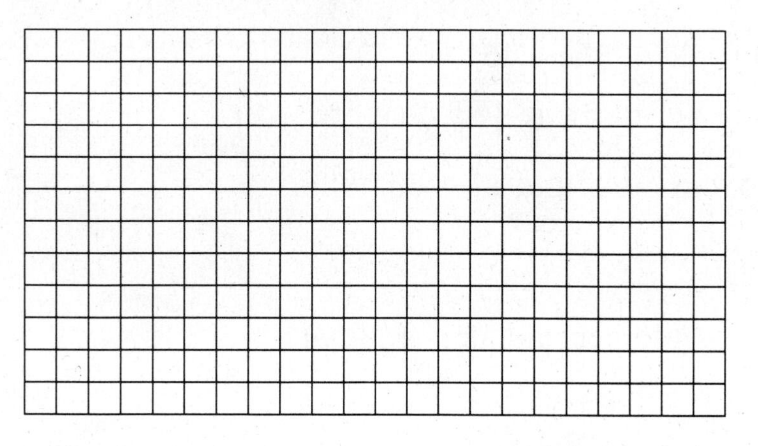

17. Every 4th day, pizzas are served on the school menus. Carrot salad is served every 6th day. If carrot salad and pizzas are served today, when is the next time they will be served together?

18. 150 baseball cards and 100 football cards are to be distributed evenly among a group of scouts. What are the possible numbers of scouts for this distribution to come out evenly?

NAME ____________________

Unit Test

1. What is the greatest common factor of 18 and 28?

A 2
B 3
C 4
D 6
E 8

2. Which of the following is a factor of 44?

A 8
B 11
C 14
D 24
E 88

3. Today Mary's aunt and grandmother came to visit. Her aunt visits every six days and her grandmother visits every eight days. The next time they will both visit Mary on the same day will be in how many days?

A 6 days
B 8 days
C 14 days
D 16 days
E 24 days

4. Tom wrote 48 as the product of two factors. One of the factors was 12. What was the other factor?

A 4
B 6
C 8
D 24
E 36

5. Which number is a multiple of 15?

A 1
B 5
C 10
D 30
E 65

NAME ______________________

Unit Test

6. Mrs. Jones invited the scouts in for apples and cookies. Each scout got the same number of apples. Each scout got the same number of cookies. Mrs. Jones served 30 apples and 40 cookies altogether. What could the number of scouts have been?

A 6
B 8
C 10
D 15
E 20

7. A set of blocks can be separated into six equal piles. It can also be separated into 15 equal piles. Which of the following is the smallest number of blocks that could be in the set?

A 3
B 21
C 30
D 45
E 90

8. Find the smallest number that has 2, 3, and 4 as factors.

A 6
B 9
C 12
D 18
E 24

9. How many different factors does 20 have?

A 2
B 3
C 4
D 5
E 6

10. Which of the following is a common multiple of 7 and 4?

A 11
B 14
C 24
D 48
E 56

NAME ____________

Unit Test

11. What is 9 squared?

A 3
B 4½
C 18
D 81
E 90

12. Which of the following is *not* a divisor of 48?

A 3
B 8
C 16
D 18
E 24

13. Which of these numbers is a composite?

A 13
B 23
C 33
D 43
E 53

14. What is always true about any prime number?

A It is divisible by only itself and 1.
B It is divisible by another prime number.
C It is less than 1,000.
D It is a factor of 1.
E It is always an odd number.

15. The four different rectangles below cover six squares on a grid. How many different rectangles can be made that would cover exactly 30 squares on a grid?

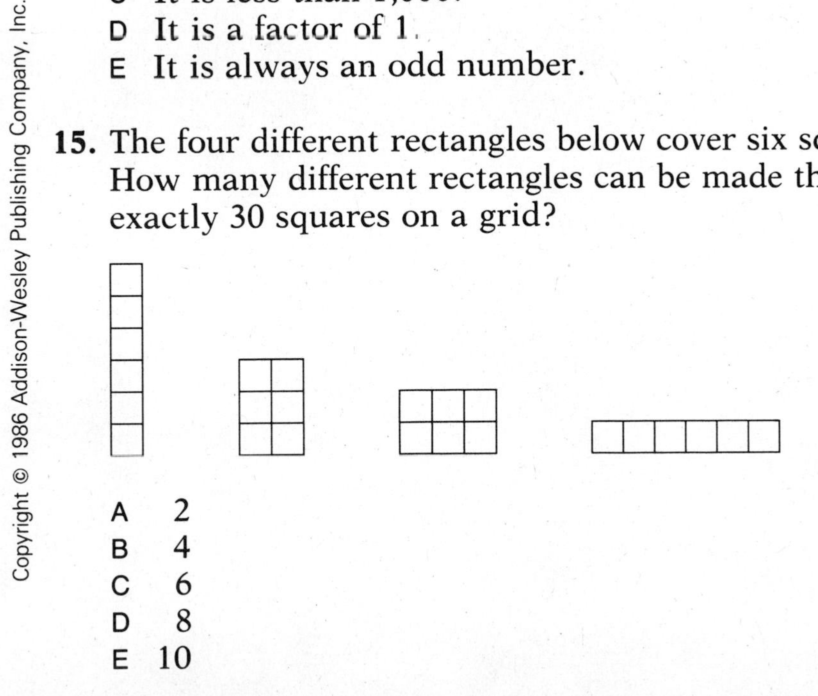

A 2
B 4
C 6
D 8
E 10

NAME ______________________

Unit Test

16. Find the smallest number that is both a multiple of 4 and a multiple of 6.

A 10
B 12
C 16
D 24
E 48

17. 3 is the greatest common divisor of 15 and another number. What could the other number be?

A 5
B 12
C 25
D 30
E 45

18. Find the prime factorization of 120.

A $3 \times 4 \times 10$
B $2 \times 3 \times 20$
C $2 \times 3 \times 4 \times 5$
D $2 \times 2 \times 2 \times 15$
E $2 \times 2 \times 2 \times 3 \times 5$

19. What is the largest factor of 250 that is less than 250?

A 25
B 50
C 75
D 125
E 200

20. John counts by 7s. Mary counts by 3s. Which of the following is a number that they both say?

A 61
B 63
C 65
D 67
E 69

NAME ______________________

Unit Test

21. The prime factorization of two numbers is $2 \times 2 \times 3 \times 5 \times 7$ and $2 \times 3 \times 5 \times 5 \times 7$. What is the greatest common factor of the two numbers?

A 2×5
B $2 \times 2 \times 5 \times 5$
C $2 \times 3 \times 5 \times 7$
D $2 \times 2 \times 3 \times 5 \times 5 \times 7$
E $2 \times 2 \times 2 \times 3 \times 3 \times 5 \times 5 \times 5 \times 7 \times 7$

22. What is the next prime after 23?

A 25
B 27
C 29
D 31
E 33

23. What is the least common multiple of 12 and 15?

A 3
B 60
C 90
D 120
E 180

24. 3, 4, and 5 are factors. What are all the possible products that can be made from a pair of these factors (the factors do not have to be different)?

A 7, 8, 9
B 6, 7, 8, 9, 10
C 8, 9, 12, 15, 20, 25
D 9, 12, 15, 16, 20, 25
E 9, 12, 15, 16, 20, 25, 60

25. How many factors does the prime factorization of 30 have?

A 2 factors
B 3 factors
C 4 factors
D 5 factors
E 6 factors

NAME

Unit Test Answer Sheet

1.	A	B	C	D	E
2.	A	B	C	D	E
3.	A	B	C	D	E
4.	A	B	C	D	E
5.	A	B	C	D	E
6.	A	B	C	D	E
7.	A	B	C	D	E
8.	A	B	C	D	E
9.	A	B	C	D	E
10.	A	B	C	D	E
11.	A	B	C	D	E
12.	A	B	C	D	E
13.	A	B	C	D	E
14.	A	B	C	D	E
15.	A	B	C	D	E
16.	A	B	C	D	E
17.	A	B	C	D	E
18.	A	B	C	D	E
19.	A	B	C	D	E
20.	A	B	C	D	E
21.	A	B	C	D	E
22.	A	B	C	D	E
23.	A	B	C	D	E
24.	A	B	C	D	E
25.	A	B	C	D	E

NAME ______________

Analyzing First Moves for the 30–Game Board

1st number picked is	Opponent Gets (Factors)	SUM	GOOD (First Moves)	NOT GOOD (First Moves)
2	1	1	2-1	
3	1	1	3-1	
4	1, 2	3	4-3	
5	1	1	5-1	
6	1, 2, 3	6	6-6	
7	1	1	7-1	
8	1, 2, 4	7	8-7	
9	1, 3	4	9-4	
10	1, 2, 5	8	10-8	
11	1	1	11-1	
12	1, 2, 3, 4, 6	16		12-16
13	1	1	13-1	
14	1, 2, 7	10	14, 10	
15	1, 3, 5	9	15-9	

NAME ______________

Analyzing First Moves for the 30–Game Board

1st number picked is	Opponent Gets (Factors)	SUM	GOOD (First Moves)	NOT GOOD (First Moves)
16	1, 2, 4, 8	15	16-15	
17	1	1	17-1	
18	1, 2, 3, 6, 9	21		18-21
19	1	1	19-1	
20	1, 2, 4, 5, 10	22		20-22
21	1, 3, 7	11	21-11	
22	1, 2, 11	14	22-14	
23	1	1	23-1	
24	1, 2, 3, 4, 6, 8, 12	36		24-36
25	1, 5	6	25-6	
26	1, 2, 13	16	26-16	
27	1, 3, 9	13	27-13	
28	1, 2, 4, 7, 14	28	28-28	
29	1	1	29-1	
30	1, 2, 3, 5, 6, 10, 15	42		30-42

NAME ____________

Moves for the 49–Game Board

Play the Factor Game on the 49–Game Board. Analyze the first moves by filling in the chart below. Since the numbers 1–30 have been done for the 30–Game Board you need only to look at the numbers 31–49.

	Factors		First Moves	
1st number picked is	Opponent Gets	SUM	GOOD	NOT GOOD
31	1	1	31-1	
32	1, 2, 4, 8, 16	31	32-31	
33	1, 3, 11	15	33-15	
34	1, 2, 17	20	34-20	
35	1, 5, 7	13	35-13	
36	1, 2, 3, 4, 6, 9, 12, 18	55		36-55
37	1	1	37-1	
38	1, 2, 19	22	38-22	
39	1, 3, 13	17	39-17	
40	1, 2, 4, 5, 8, 10, 20	50		40-50
41	1	1	41-1	
42	1, 2, 3, 6, 7, 14, 21	54		42-54
43	1	1	43-1	
44	1, 2, 4, 11, 22	40	44-40	
45	1, 3, 5, 9, 15	33	45-33	
46	1, 2, 23	26	46-26	
47	1	1	47-1	
48	1, 2, 3, 4, 6, 8, 12, 16, 24	76		48-76
49	1, 7	8	49-8	

Worksheet 1-3 18

NAME ____________

Practice Problems

Use the table of first moves for numbers 1 through 49 (Worksheet 1-3) to complete the following.

1. List all the prime numbers from 1 to 49.

2, 3, 5, 7, 11, 13, 17, 19, 23, 29, 31, 37, 41, 43, 47

2. List all the numbers from 1 to 49 that are abundant numbers.

12, 18, 20, 24, 30, 36, 40, 42, 48

3. List all the numbers from 1 to 49 that are deficient numbers.

The primes plus 4, 8, 9, 10, 14, 15, 16, 21, 22, 25, 26, 27, 32, 33, 34, 35, 38, 39, 44, 45, 46, 49

4. List all the numbers from 1 to 49 that are perfect numbers.

6, 28

5. List all the numbers that have 2 as a factor.

2, 4, 6, 8, 10, 12, 14, 16, 18, 20, 22, 24, 26, 28, 30, 32, 34, 36, 38, 40, 42, 44, 46, 48

What do we call these numbers? even numbers, or multiples of two

What do we call numbers that do not have 2 as a factor? odd numbers

Worksheet 1-4 19

NAME

Practice Problems

6. What number(s) have the most factors? 48

7. What factor is paired with
16 to give 48? 3 16 to give 32? 2

8. What factor is paired with
12 to give 48? 4 12 to give 36? 3

9. What factor is paired with
6 to give 48? 8 6 to give 42? 7

10. What factor is paired with
4 to give 48? 12 4 to give 44? 11

11. What factor is paired with
2 to give 48? 24 2 to give 34? 17

12. What is the *best* first move on a 49–Game Board?
47

13. What is the *worst* first move on a 49–Game Board?
48

20 Worksheet 1-4, page 2

NAME

Create a 3 × 3 Product Game

1	2	3
4	6	8
9	12	16

3 × 3
Factors:
1, 2, 3, 4

Factors	Possible Products
1	1
2	2, 4
3	3, 6, 9
4	8, 12, 16

30 Worksheet 2-2

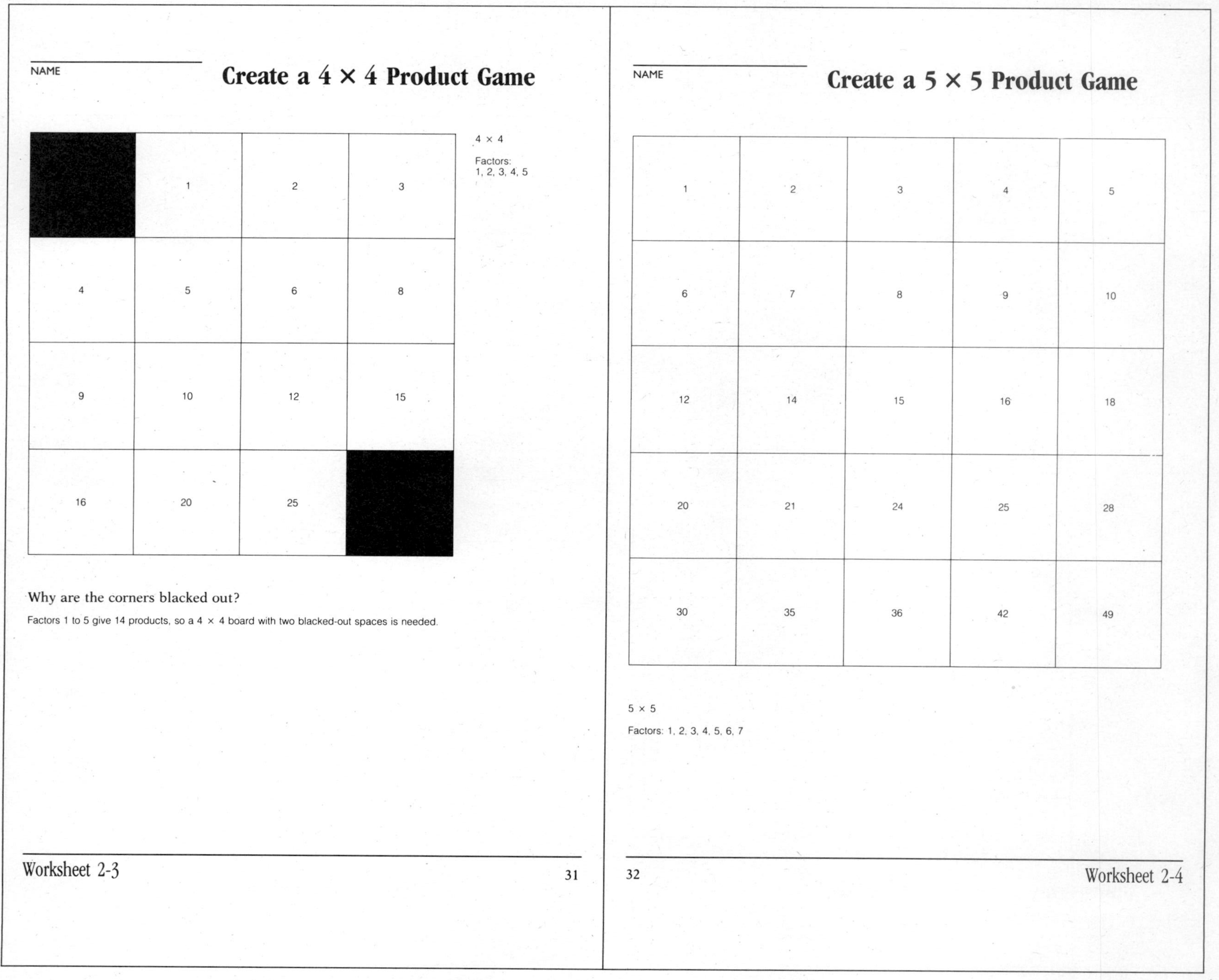

NAME

Create a 4 × 4 Product Game

■	1	2	3
4	5	6	8
9	10	12	15
16	20	25	■

4 × 4
Factors:
1, 2, 3, 4, 5

Why are the corners blacked out?

Factors 1 to 5 give 14 products, so a 4 × 4 board with two blacked-out spaces is needed.

Worksheet 2-3 31

NAME

Create a 5 × 5 Product Game

1	2	3	4	5
6	7	8	9	10
12	14	15	16	18
20	21	24	25	28
30	35	36	42	49

5 × 5
Factors: 1, 2, 3, 4, 5, 6, 7

32 Worksheet 2-4

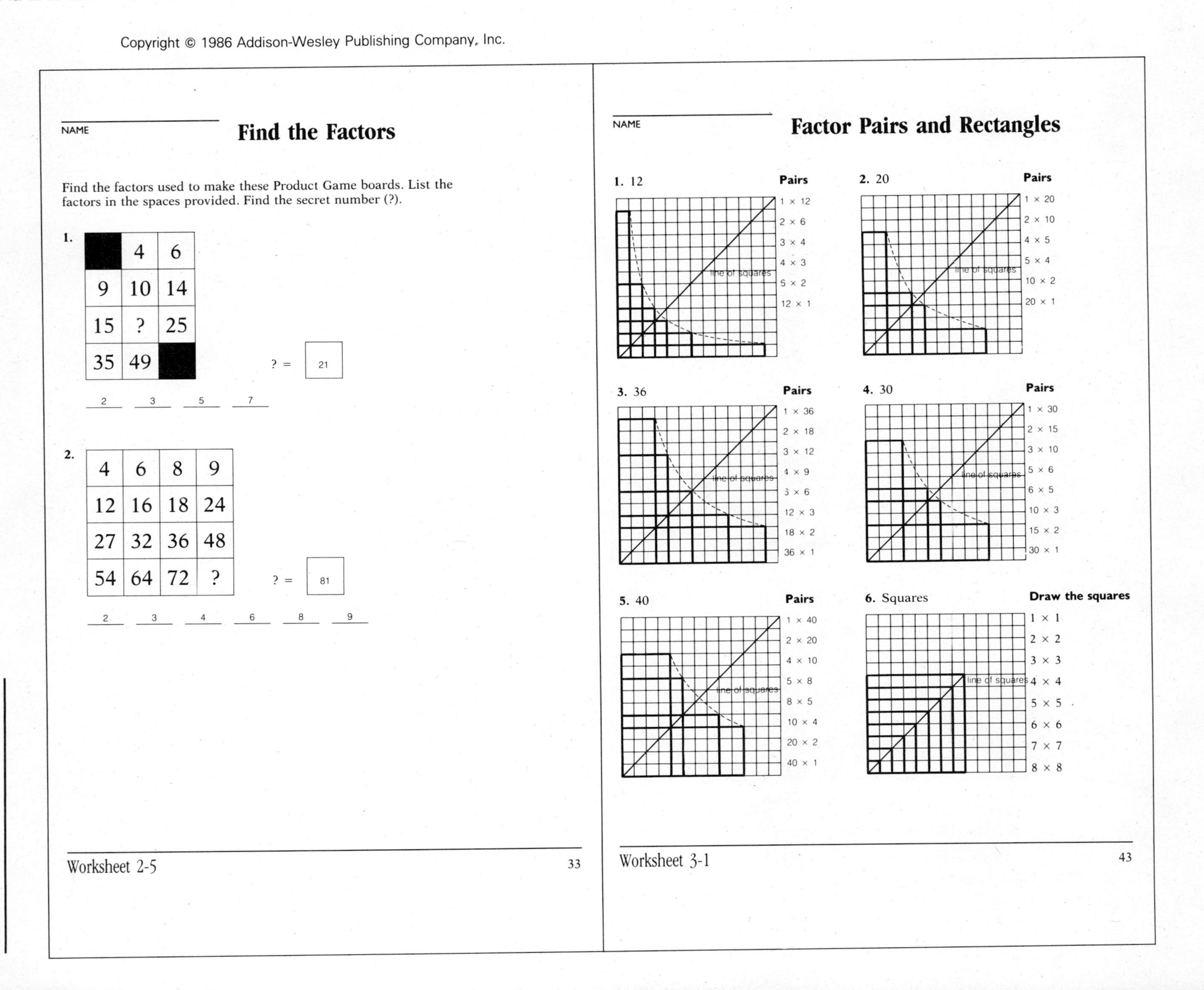

NAME ____________

Find the Factors

Find the factors used to make these Product Game boards. List the factors in the spaces provided. Find the secret number (?).

1.

■	4	6
9	10	14
15	?	25
35	49	■

? = 21

2, 3, 5, 7

2.

4	6	8	9
12	16	18	24
27	32	36	48
54	64	72	?

? = 81

2, 3, 4, 6, 8, 9

Worksheet 2-5 33

NAME ____________

Factor Pairs and Rectangles

1. 12

Pairs
1 × 12
2 × 6
3 × 4
4 × 3
5 × 2
12 × 1

2. 20

Pairs
1 × 20
2 × 10
4 × 5
5 × 4
10 × 2
20 × 1

3. 36

Pairs
1 × 36
2 × 18
3 × 12
4 × 9
3 × 6
12 × 3
18 × 2
36 × 1

4. 30

Pairs
1 × 30
2 × 15
3 × 10
5 × 6
6 × 5
10 × 3
15 × 2
30 × 1

5. 40

Pairs
1 × 40
2 × 20
4 × 10
5 × 8
8 × 5
10 × 4
20 × 2
40 × 1

6. Squares

Draw the squares
1 × 1
2 × 2
3 × 3
4 × 4
5 × 5
6 × 6
7 × 7
8 × 8

Worksheet 3-1 43

NAME

Crossing the Line

For each of the numbers, find where the factor pairs cross the line of squares.

1. 22

between 4 and 5

2. 44

between 6 and 7

3. 64

at 8

4. 128

between 11 and 12

5. 154

between 12 and 13

6. 278

between 16 and 17

7. To find all the factor pairs for 300, how far do you have to check to be sure that you have found them all?

all numbers less than 17

8. List all the factor pairs for 300 (including the reversed ones).

1 × 300	6 × 50	25 × 12	75 × 4
2 × 150	10 × 30	30 × 10	100 × 3
3 × 100	12 × 25	50 × 6	150 × 2
4 × 75	15 × 20	60 × 5	300 × 1
5 × 60	20 × 15		

NAME

Practice Problems

1. Mr. Brown wrote a number on the blackboard and said, "I know that to be sure I find all the factor pairs for this number, I must check each number from 1 to 47."

a) What is the smallest number Mr. Brown could have written on the blackboard?

47 × 47 = 2,209

b) What is the largest number Mr. Brown could have written on the blackboard?

2,304 − 1 = 2,303 $(48^2 = 2{,}304)$

2. Bob is making factor pairs of a number. He finds he can make exactly 7 factor pairs (including the reversed ones). If his number is less than 100, what is it?

64

1 × 64
2 × 32
4 × 16
8 × 8
16 × 4
32 × 2
64 × 1

3. Jo has chosen a number larger than 12 and less than 50. She finds exactly 6 factor pairs (including the reversed ones). What could her number be?

18, 32, 45, 28, or 20

1 × 18	1 × 32	1 × 45	1 × 28	1 × 20
2 × 9	2 × 16	3 × 15	2 × 14	2 × 10
3 × 6	4 × 8	5 × 9	4 × 7	4 × 5
6 × 3	8 × 4	9 × 5	7 × 4	5 × 4
9 × 2	16 × 2	15 × 3	14 × 2	10 × 2
18 × 1	32 × 1	45 × 1	28 × 1	20 × 1

NAME ____________

Product Puzzle I

Draw a loop around strings of numbers whose product is 1,350.

Answers will vary; possibilities are illustrated below.

5	×	6	×	5	×	9	×	54
×	3	×	3	×	150	×	9	×
10	×	27	×	5	×	5	×	2
×	54	×	25	×	3	×	45	×
9	×	5	×	6	×	5	×	135
×	5	×	9	×	150	×	2	×
15	×	10	×	9	×	3	×	5
×	5	×	6	×	45	×	15	×
1	×	3	×	25	×	9	×	2

Record the different strings you have found.

1. 5 × 6 × 5 × 9
2. 150 × 9
3. 10 × 27 × 5
4. 6 × 9 × 25
5. 5 × 2 × 3 × 45
6. 54 × 25
7. 2 × 45 × 15
8. and so on
9. ____________
10. ____________
11. ____________
12. ____________
13. ____________
14. ____________
15. ____________

After you have found 15 different strings, answer these questions.

1. Find 2 strings of numbers whose product is 1,350 that are not in the table.

   Answers will vary; possible answers include 1,350 × 1; 75 × 2 × 9; 2 × 675.

2. Can you find a string whose product is 1,350 that is longer than any in the table?

   2 × 3 × 3 × 3 × 5 × 5

Worksheet 4-1 55

NAME ____________

Factor Trees I

For each of these numbers, grow a factor tree. When you reach the bottom of a tree, check it by multiplying the numbers together.

1. 36
   2 × 18
   2 × 2 × 9
   2 × 2 × 3 × 3

2. 84
   2 × 42
   2 × 6 × 7
   2 × 2 × 3 × 7

3. 72
   2 × 36
   2 × 2 × 2 × 3 × 3

4. 57
   3 × 19

5. 144
   2 × 72
   2 × 2 × 2 × 2 × 3 × 3

6. 147
   3 × 49
   3 × 7 × 7

7. 63
   3 × 21
   3 × 3 × 7

8. 64
   8 × 8
   2 × 2 × 2 × 2 × 2 × 2

56 Worksheet 4-2

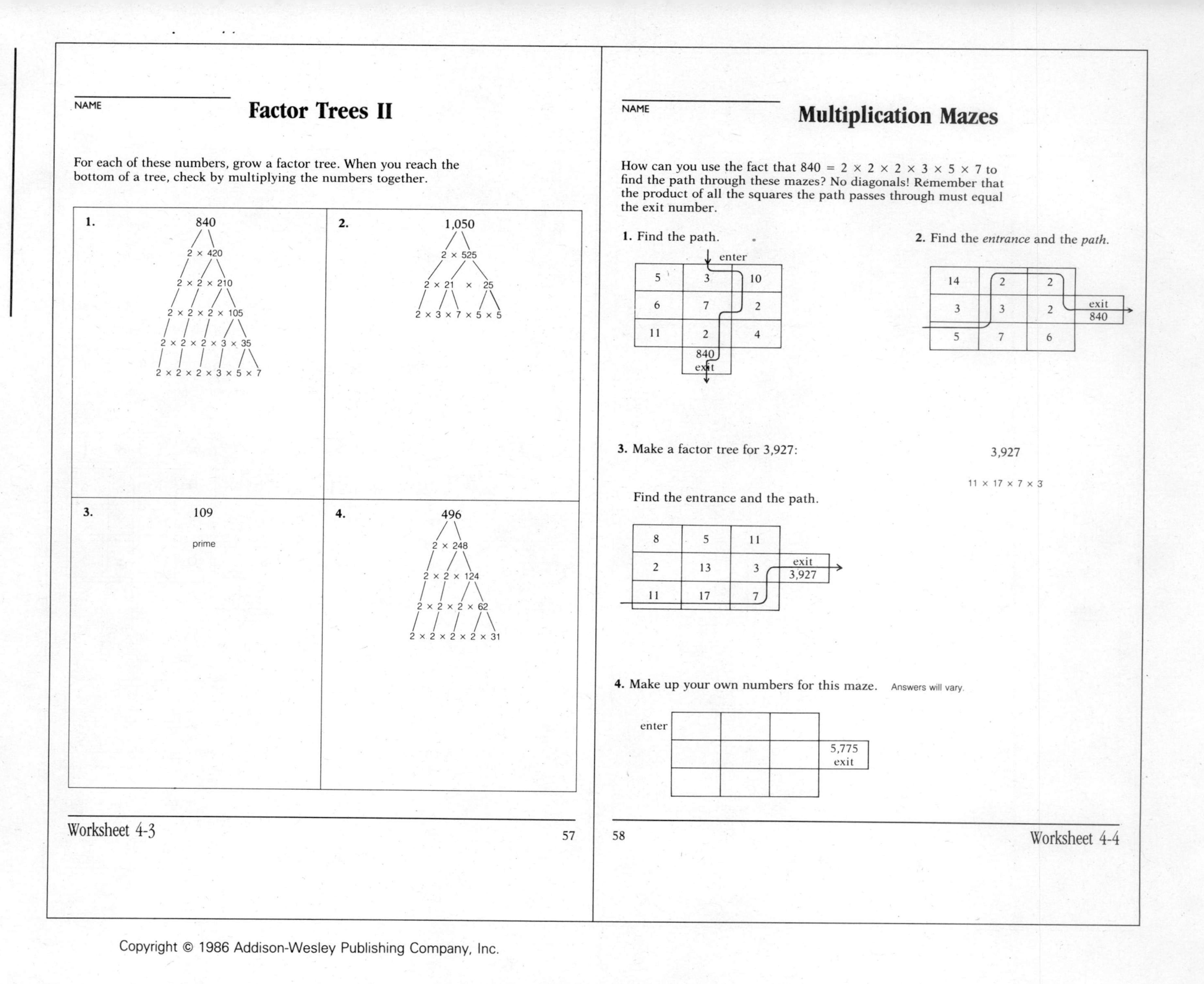

NAME

Factor Trees II

For each of these numbers, grow a factor tree. When you reach the bottom of a tree, check by multiplying the numbers together.

1. 840

2 × 420

2 × 2 × 210

2 × 2 × 2 × 105

2 × 2 × 2 × 3 × 35

2 × 2 × 2 × 3 × 5 × 7

2. 1,050

2 × 525

2 × 21 × 25

2 × 3 × 7 × 5 × 5

3. 109

prime

4. 496

2 × 248

2 × 2 × 124

2 × 2 × 2 × 62

2 × 2 × 2 × 2 × 31

Worksheet 4-3 57

NAME

Multiplication Mazes

How can you use the fact that 840 = 2 × 2 × 2 × 3 × 5 × 7 to find the path through these mazes? No diagonals! Remember that the product of all the squares the path passes through must equal the exit number.

1. Find the path.

enter

5	3	10
6	7	2
11	2	4

840 exit

2. Find the *entrance* and the *path*.

14	2	2
3	3	2
5	7	6

exit 840

3. Make a factor tree for 3,927:

3,927

11 × 17 × 7 × 3

Find the entrance and the path.

8	5	11
2	13	3
11	17	7

exit 3,927

4. Make up your own numbers for this maze. Answers will vary.

enter

5,775 exit

58 Worksheet 4-4

NAME

Product Puzzle II

Find all strings whose product is 630. Strings may go horizontally, vertically, or bend around corners.

7	×	3	×	30
×	5	×	42	×
6	×	15	×	21
×	45	×	9	×
5	×	14	×	2

7 × 3 × 30

3 × 7 × 6 × 5

21 × 30

15 × 21 × 2

9 × 14 × 5

15 × 42

3 × 42 × 5

3 × 15 × 14

NAME

Least Common Multiples

In problems 1–3, use a calculator to find the multiples of each number. Write the least common multiple in the box. Show what each of the numbers must be multiplied by to get the LCM.

1. Multiples of 3: 3, 6, 9, 12, 15, 18, 21, 24

Multiples of 4: 4, 8, 12, 16, 20, 24, 28, 32

Common multiples of 3 and 4. 12, 24 . . .

3 × 4
4 × 3 = LCM = 12

2. Multiples of 30: 30, 60, 90, 120, 150, 180, 210, 240

Multiples of 42: 42, 84, 126, 168, 210, 252, 294, 336

Common multiples of 30 and 42: 210 . . .

30 × 7
42 × 5 = LCM = 210

3. Multiples of 18: 18, 36, 54, 72, 90, 108, 126, 144

Multiples of 36: 36, 72, 108, 144, 180, 216, 252, 288

Common multiples of 18 and 36: 36, 72, 108, 144 . . .

18 × 2
36 × 1 = LCM = 36

In problems 4–6, make a factor tree for each of the numbers. Use the prime factorizations to find the least common multiple.

4. 3 — prime; 4 — 2 × 2

String for LCM: 3 × 2 × 2 = 12

5. 30 — 2 × 15 — 2 × 3 × 5; 40 — 2 × 20 — 2 × 2 × 2 × 5

String for LCM: 2 × 3 × 5 × 2 × 2 = 120

6. 18 — 2 × 9 — 2 × 3 × 3; 35 — 5 × 7

String for LCM: 2 × 3 × 3 × 5 × 7 = 630

NAME

Practice Problems

Find the least common multiple (LCM) of each pair of numbers given.

1. 4, 9 LCM is 36

2. 8, 14 LCM is 56

3. 10, 45 LCM is 90

4. 14, 15 LCM is 210

5. 14, 21 LCM is 42

6. 24, 36 LCM is 72

7. 58, 96 LCM is 2,784

8. 180, 210 LCM is 1,260

Find all the common multiples less than 100 for each pair of numbers given.

9. Common multiples of 5 and 7 35, 70

10. Common multiples of 12 and 20 60

11. Common multiples of 4 and 6 12, 24, 36, 48, 60, 72, 84, 96

12. Common multiples of 9 and 15 45, 90

What is the smallest number that has the following factors?

13. 2 and 3. 6

14. 2, 3 and 4. 12

15. 4, 6 and 9. 36

16. 4, 6, 9 and 12. 36

17. I am thinking of a number. The least common multiple of my number and 9 is 45. What could my number be?

5, 15, or 45

18. I am thinking of a number. My number has 8 as a factor and 12 as a factor.

a) What is the smallest that my number could be? 24

b) Name four other numbers that are factors of my number. 2, 3, 4, 6

19. a) Find the LCM of 12 and 35.

420

b) Name 3 other common multiples of 12 and 35.

840, 1,260, 1,680

c) Name 3 other factors of the LCM.

2, 3, 5, 7, 4, 6, 10, 14, 20, 28, 30, 42, 60, 84, 105, 210

NAME

Applications of LCM

1. Gleamy-Tooth toothpaste comes in 2 sizes.

GLEAMY-TOOTH 9 oz for $0.89

GLEAMY-TOOTH 12 oz for $1.19

a) What is the LCM of 9 and 12? 36

b) If you bought that much toothpaste in 9-oz tubes, how much would it cost? $3.56

c) If you bought that much toothpaste in 12-oz tubes, how much would it cost? $3.57

d) Which tube gives you more Gleamy-Tooth for the money? 9 oz for $0.89

2. In the school kitchen during lunch, the timer for pizza buzzes every 14 minutes; the timer for hamburger buns buzzes every 6 minutes. The two timers just buzzed together. In how many minutes will they buzz together again? 42 minutes

3. Two ships sail steadily between New York and London. One ship takes 12 days to make a round trip; the other takes 15 days. If they are both in New York today, in how many days will they both be in New York again? 60 days

4. The high school lunch menu repeats every 20 days; the elementary school menu repeats every 15 days. Both schools are serving sloppy joes today. In how many days will they both serve sloppy joes again? 60 days

5. Two neon signs are turned on at the same time. One blinks every 4 seconds; the other blinks every 6 seconds. How many times per minute do they blink on together? at 12, 24, 36, 48, and 60 seconds or 5 times per minute

6. How many teeth should be on gear A if each turn of gear A is to produce a whole number of turns of the shafts attached to B and C?

36 teeth

B
C
A

B has 12 teeth
C has 18 teeth

NAME

Common Factors

List the factors of each number. Then find the common factors of the two numbers. Do all your work in the space provided.

1. 9 and 24
1, 3

2. 32 and 48
1, 2, 4, 8, 16

3. 51 and 17
1, 17

4. 52 and 8
1, 2, 4

5. 1,001 and 70
1, 7

6. 56 and 35
1, 7

7. The scout leader has a certain number of cookies. They can be divided evenly among 9 scouts. They can also be divided evenly among 6 scouts. What are *two* possibilities for the number of cookies? 36, 18

other possibilities: 54, 72 (multiples of 18)

8. A band of pirates divided 185 pieces of silver and 148 gold coins. These pirates were known to be absolutely fair about sharing equally. How many pirates were there? 37

NAME

Greatest Common Factor

Solve problems 1–4 by finding the prime factorization of each number. Circle the largest string that is in both numbers. Then write the value of the greatest common factor in the box.

1. 56 (2 × 2) × 2 × 7
36 (2 × 2) × 3 × 3
GCF [4]

2. 18 (2 × 3) × 3
24 2 × (2 × 3)
GCF [6]

3. 36 2 × 2 × 3 × 3
19 19
GCF [1]

4. 52 2 × 2 × (13)
13 (13)
GCF [13]

Use the following story to solve problems 5 and 6.

Ms. Wurst and Mr. Pop have donated a total of 91 hotdogs and 126 small cans of fruit juice for a math class picnic. Each student will receive the same amount of refreshments.

5. What is the greatest number of students that can attend the picnic? 7

How many cans of juice will each student receive? 18

How many hotdogs will each student receive? 13

6. If one of the hotdogs is eaten by Ms. Wurst's dog just before the picnic, what is the greatest number of students that can attend? 18

How many hotdogs will each student receive? 5

How many cans of juice will each student receive? 7

NAME ______

Practice Problems

List all the factors that the following sets of numbers have in common.

1. 21 and 49 1, 7

2. 17 and 37 1

3. 12, 36, and 48 1, 2, 3, 4, 6, 12

4. 92 and 180 1, 2, 4

What is the greatest common factor (GCF) of each of the following numbers?

5. 18 and 36 18

6. 29 and 49 1

7. 165 and 198 33

8. 630 and 1,350 90

9. What is the smallest number that 8 and 12 both divide? 24

10. a) If 8 and 20 both divide a number N, name four other numbers that must divide N. 1, 2, 4, 5, 10, 40

b) What is the smallest number that could be N? 40

Worksheet 6-3 85

NAME ______

100 Board

Code

2	3
7	5

Multiples of 2, 3, 5, and 7 are sifted.

Primes from 1 to 100: 2, 3, 5, 7, 11, 13, 17, 19, 23, 29, 31, 37, 41, 43, 47, 53, 59, 61, 67, 71, 73, 79, 83, 89, 97

1	2	3	4	5	6	7	8	9	10
11	12	13	14	15	16	17	18	19	20
21	22	23	24	25	26	27	28	29	30
31	32	33	34	35	36	37	38	39	40
41	42	43	44	45	46	47	48	49	50
51	52	53	54	55	56	57	58	59	60
61	62	63	64	65	66	67	68	69	70
71	72	73	74	75	76	77	78	79	80
81	82	83	84	85	86	87	88	89	90
91	92	93	94	95	96	97	98	99	100

Worksheet 7-1 97

NAME

101–200 Board

Code

2	3
13	11
7	5

Multiples of 2, 3, 5, 7, 11, and 13 are sifted.

Primes from 100 to 200: 101, 103, 107, 109, 113, 127, 131, 137, 139, 149, 151, 157, 163, 167, 173, 179, 181, 191, 193, 197, 199

101	102	103	104	105	106	107	108	109	110
111	112	113	114	115	116	117	118	119	120
121	122	123	124	125	126	127	128	129	130
131	132	133	134	135	136	137	138	139	140
141	142	143	144	145	146	147	148	149	150
151	152	153	154	155	156	157	158	159	160
161	162	163	164	165	166	167	168	169	170
171	172	173	174	175	176	177	178	179	180
181	182	183	184	185	186	187	188	189	190
191	192	193	194	195	196	197	198	199	200

NAME

Using the Sieve

Use your 100 Board to answer the following questions.

1. What is the smallest prime number that is greater than 30?

   31

2. What is the smallest prime number that is greater than 50?

   53

3. 5 and 7 are called *twin primes* because they are both primes and they differ by two. List all twin primes between 1 and 100.

   11, 13; 17, 19; 29, 31; 41, 43; 71, 73

4. A number that is not prime (and not 1) is a composite number. Find 5 composite numbers in a row.

   24–28; 32–36; 48–52; 54–58; 62–66; 74–78; 84–88. 90–96 is a string of 6 composite numbers.

5. Why didn't we sift for 9s?

   All multiples of 9 are also multiples of 3 and are sifted when we sift the 3s.

6. Which of the primes 2, 3, 5, and 7 divide 84?

   2, 3, 7

7. The number 6 was sifted with both 2 and 3.

   a) Find all other numbers that were sifted with both 2 and 3.

   6, 12, 18, 24, 30, 36, 42, 48, 54, 60, 66, 72, 78, 84, 90, 96

   b) How are all the multiples of 6 marked on the board?

   2/3 and possibly other lines

8. **a)** List the multiples of 7.

   7, 14, 21, 28, 35, 42, 49, 56, 63, 70, 77, 84, 91, 98

   b) What numbers are multiples of *both* 6 and 7.

   42, 84

9. How are multiples of 15 marked on the board?

   3/5 and possibly other lines

10. There are four columns on the board that contain no primes. Find them and explain why these columns contain no primes.

    The columns headed headed by 4, 6, 8, and 10 contain only even numbers. They are all sifted as multiples of 2.

NAME

Prime Puzzle

There is a message hidden below. Cross out the letters in the boxes containing numbers that are *not* prime numbers to discover the message in the remaining boxes.

D 7	~~P 6~~	I 2	~~R 8~~	V 19	I 11	~~M 12~~	~~P 60~~	S 3	~~K 9~~	~~S 14~~	O 59	~~Z 35~~	R 11	S 37
~~O 4~~	A 3	R 31	~~M 25~~	E 23	~~S 10~~	D 29	~~M 12~~	I 41	V 97	~~H 100~~	I 23	N 83	E 13	~~A 12~~
B 71	U 2	~~R 35~~	T 3	~~T 27~~	F 43	~~Q 42~~	A 37	~~I 64~~	C 7	T 5	~~R 45~~	O 13	R 11	S 71
~~N 9~~	~~E 14~~	~~U 69~~	~~M 32~~	A 17	~~S 87~~	~~F 48~~	~~G 75~~	~~Q 20~~	R 19	~~K 9~~	E 97	~~Q 8~~	~~T 27~~	~~D 57~~
F 67	R 2	~~C 16~~	I 89	~~M 18~~	E 7	~~T 12~~	~~K 9~~	N 17	D 73	L 67	~~N 49~~	I 59	E 29	R 83

Divisors are divine but factors are friendlier.

NAME

Practice Problems

Use your 100–200 Board to answer the following questions.

1. List all the primes between 100 and 200.

101, 103, 107, 109, 113, 127, 131, 137, 139, 149, 151, 157, 163, 167, 173, 179, 181, 191, 193, 197, 199

2. What prime numbers are factors of 84? 2, 3, 7

3. What prime numbers are factors of 110? 2, 5, 11

4. List all the multiples of 7 that are between 130 and 200.

133, 140, 147, 154, 161, 168, 175, 182, 189, 196

5. Why didn't we sift for 17s on the 200 board?

The first unsifted multiple of 17 would be $17 \times 17 = 289$. This is not on the board.

6. A number that is not prime (and not 1) is a composite number. Find 5 composite numbers in a row on the 200 board. What is the longest string you can find?

132–136; 152–156; 158–162; 168–172; and 174–178 are strings of 5 composites. 114–127 is a string of 13 composites. 140–148 is a string of 9 composites. 182–190 is a string of 9 composites.

7. Why is it impossible to find a number on the 200 board that is divisible by four different prime numbers? (Hint: How big would such a number have to be?)

$2 \times 3 \times 5 \times 7 = 210$

8. What would be the last prime we would have to sift to find all primes less than 1,000?

31

Answers

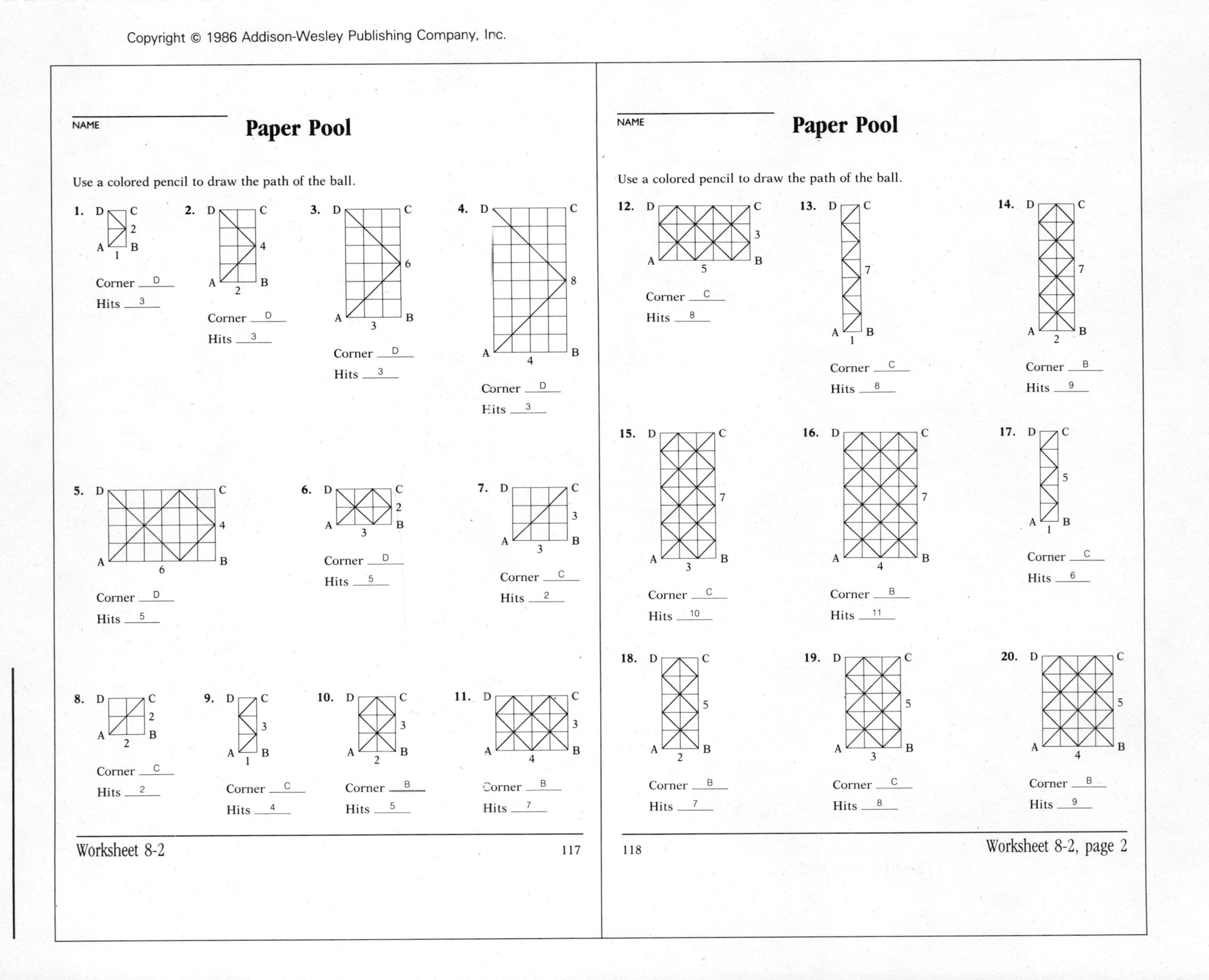

NAME
Paper Pool
Use a colored pencil to draw the path of the ball.
1. Corner D Hits 3
2. Corner D Hits 3
3. Corner D Hits 3
4. Corner D Hits 3
5. Corner D Hits 5
6. Corner D Hits 5
7. Corner C Hits 2
8. Corner C Hits 2
9. Corner C Hits 4
10. Corner B Hits 5
11. Corner B Hits 7
Worksheet 8-2
117
NAME
Paper Pool
Use a colored pencil to draw the path of the ball.
12. Corner C Hits 8
13. Corner C Hits 8
14. Corner B Hits 9
15. Corner C Hits 10
16. Corner B Hits 11
17. Corner C Hits 6
18. Corner B Hits 7
19. Corner C Hits 8
20. Corner B Hits 9
118
Worksheet 8-2, page 2

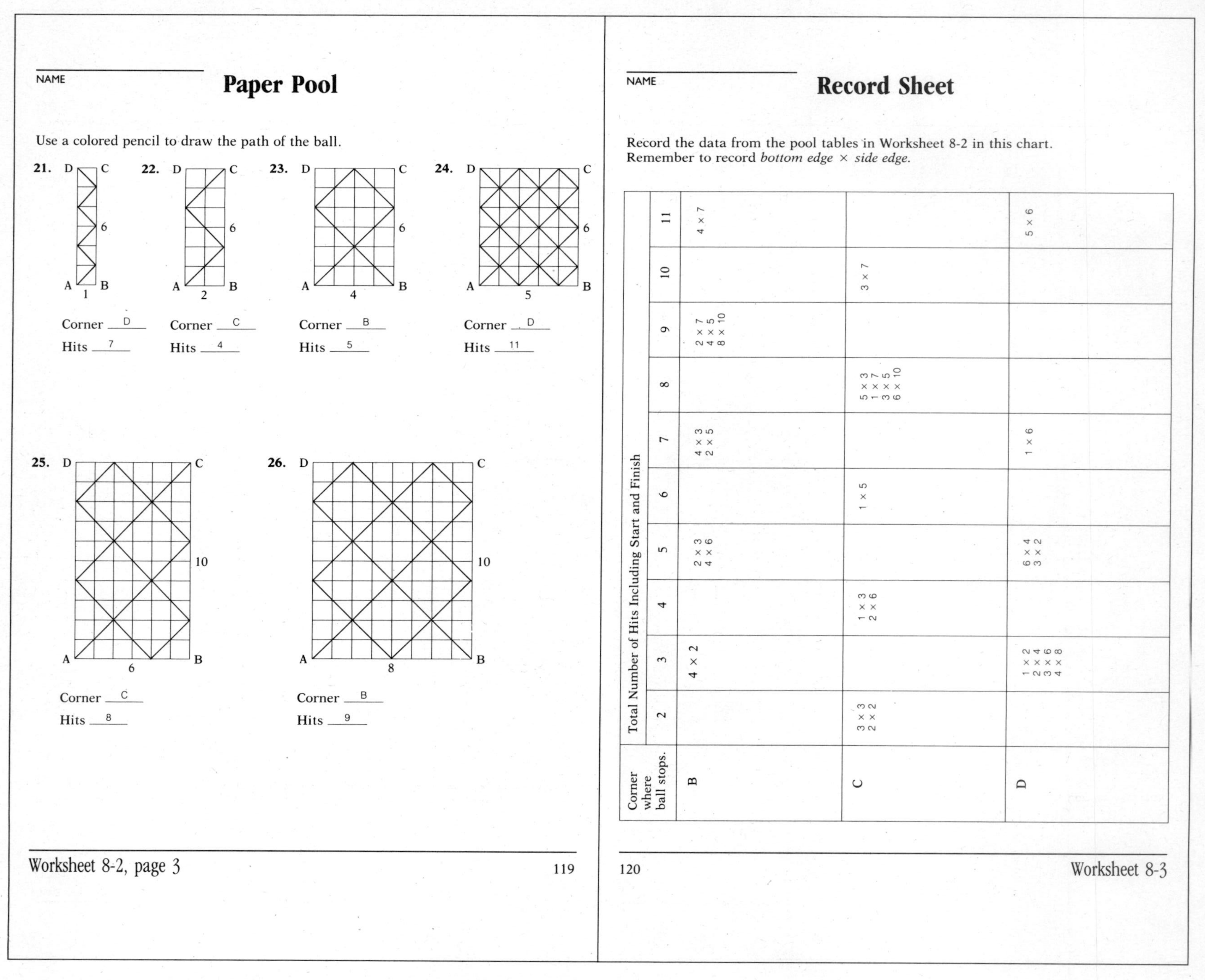

NAME

Paper Pool

Use a colored pencil to draw the path of the ball.

21. Corner D Hits 7

22. Corner C Hits 4

23. Corner B Hits 5

24. Corner D Hits 11

25. Corner C Hits 8

26. Corner B Hits 9

Worksheet 8-2, page 3 119

NAME

Record Sheet

Record the data from the pool tables in Worksheet 8-2 in this chart.
Remember to record *bottom edge* × *side edge*.

Corner where ball stops.	Total Number of Hits Including Start and Finish: 2	3	4	5	6	7	8	9	10	11
B		4 × 2		2 × 3 4 × 6		4 × 3 2 × 5		2 × 7 4 × 5 8 × 10		4 × 7
C	3 × 3 2 × 2		1 × 3 2 × 6		1 × 5		5 × 3 1 × 7 3 × 5 6 × 10		3 × 7	
D		1 × 2 2 × 4 3 × 6 4 × 8		6 × 4 3 × 2		1 × 6				5 × 6

120 Worksheet 8-3

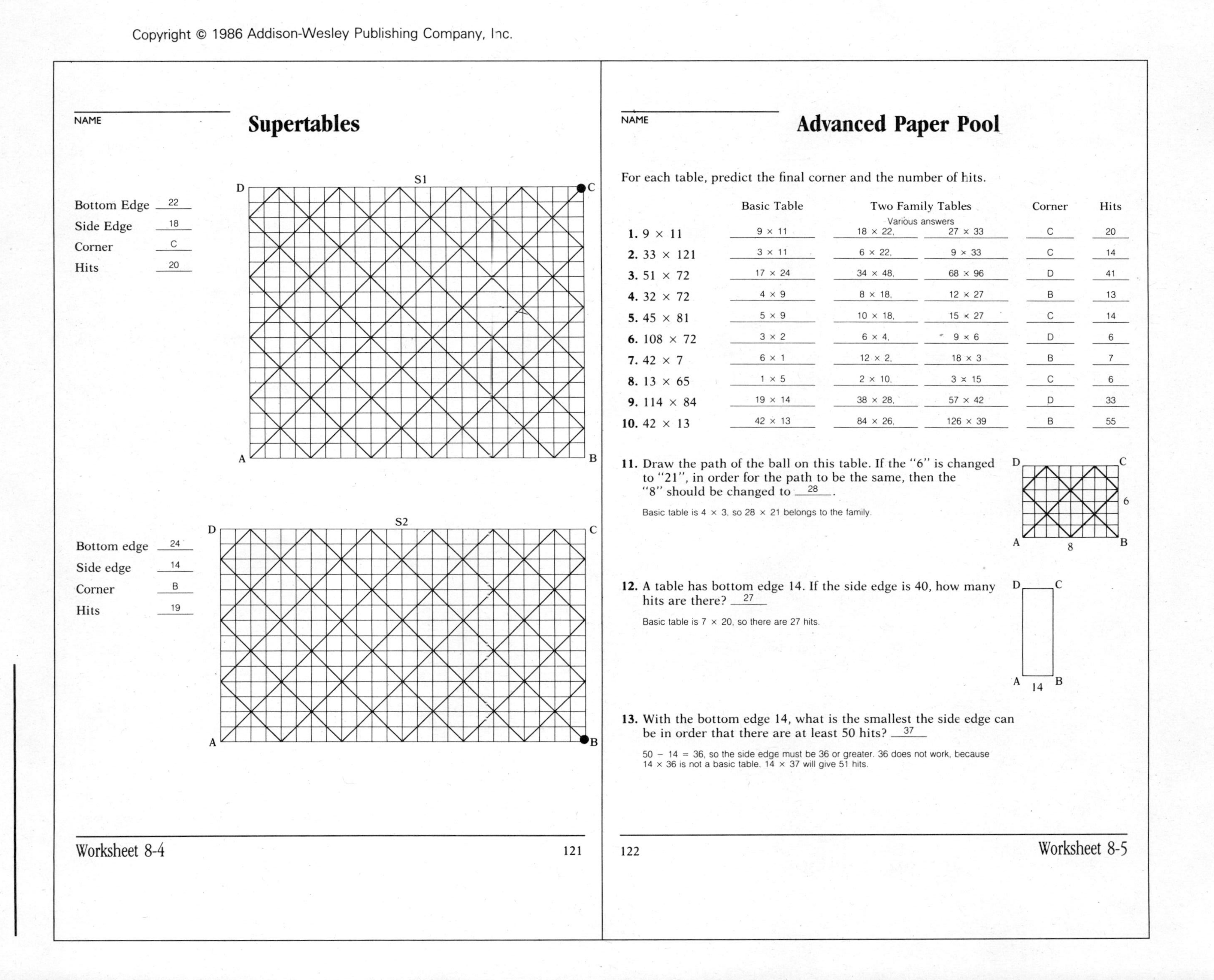

NAME

Supertables

Bottom Edge	22
Side Edge	18
Corner	C
Hits	20

Bottom edge	24
Side edge	14
Corner	B
Hits	19

NAME

Advanced Paper Pool

For each table, predict the final corner and the number of hits.

	Basic Table	Two Family Tables (Various answers)		Corner	Hits
1. 9 × 11	9 × 11	18 × 22,	27 × 33	C	20
2. 33 × 121	3 × 11	6 × 22,	9 × 33	C	14
3. 51 × 72	17 × 24	34 × 48,	68 × 96	D	41
4. 32 × 72	4 × 9	8 × 18,	12 × 27	B	13
5. 45 × 81	5 × 9	10 × 18,	15 × 27	C	14
6. 108 × 72	3 × 2	6 × 4,	9 × 6	D	6
7. 42 × 7	6 × 1	12 × 2,	18 × 3	B	7
8. 13 × 65	1 × 5	2 × 10,	3 × 15	C	6
9. 114 × 84	19 × 14	38 × 28,	57 × 42	D	33
10. 42 × 13	42 × 13	84 × 26,	126 × 39	B	55

11. Draw the path of the ball on this table. If the "6" is changed to "21", in order for the path to be the same, then the "8" should be changed to 28.

Basic table is 4 × 3, so 28 × 21 belongs to the family.

12. A table has bottom edge 14. If the side edge is 40, how many hits are there? 27

Basic table is 7 × 20, so there are 27 hits.

13. With the bottom edge 14, what is the smallest the side edge can be in order that there are at least 50 hits? 37

50 − 14 = 36, so the side edge must be 36 or greater. 36 does not work, because 14 × 36 is not a basic table. 14 × 37 will give 51 hits.

NAME

How Many Squares Are Crossed?

There is also a rule to predict how many squares the path crosses on a paper pool table.

For example: This path crosses two squares. Use your Paper Pool worksheets and record the data from each table.

Size of Table	Number of Squares Crossed
* 1 × 2	2
2 × 4	4
3 × 6	6
4 × 8	8
6 × 4	12
* 3 × 2	6
3 × 3	3
2 × 2	2
* 1 × 3	3
* 2 × 3	6
* 4 × 3	12
and so on	

Size of Table	Number of Squares Crossed
* 1 × 7	7
* 2 × 7	14
* 3 × 7	21
* 4 × 7	28
* 1 × 5	5
* 2 × 5	10
* 3 × 5	15
* 4 × 5	20
* 1 × 6	6
2 × 6	6
3 × 6	6
and so on	

Mark the basic tables with a *.

What is the rule for the basic tables? Product of bottom edge × side edge

What is the rule for the other tables? least common multiple of the bottom edge and side edge

Does the rule for the other tables also work for basic tables?
Yes. If they are basic tables, then the dimensions are relatively prime; therefore the LCM is simply the product.

How many squares would be crossed in a 24 × 36 table? 72

NAME

Review Problems

1. Find all the factor pairs for the following numbers.

a) 28
1 × 28 28 × 1
2 × 14 14 × 2
4 × 7 7 × 4

b) 72
1 × 72 72 × 1
2 × 36 36 × 2
3 × 24 24 × 3
4 × 18 18 × 4
6 × 12 12 × 6

2. What is the smallest number we have to check to be sure that we have all the factor pairs of the following numbers?

a) 60 7
b) 150 12

3. Decide whether the following numbers are deficient, perfect, or abundant.

a) 12 abundant
b) 28 perfect
c) 21 deficient
d) 64 deficient

4. Fill in the box with a number that will make each statement true.

a) 7 × [8] = 56

b) [12] × 3 = 36

c) 14 × 5 = [70]

5. Find the factors used to make the following Product Game.

4	■	6	9
10	14	15	■
21	25	35	49

2, 3, 5, 7

6. Find all the divisors of the following numbers.

a) 30 1, 2, 3, 5, 6, 10, 15, 30
b) 29 1, 29

NAME

Review Problems

7. Mrs. Brown wrote a number on the chalkboard and said, "I know that to be sure I find all the factor pairs for this number I must check each number from 1–29." What numbers could Mrs. Brown have written? What is the smallest number Mrs. Brown could have written? What is the largest number Mrs. Brown could have written?

All the numbers from 29^2 through $30^2 - 1$.

$29^2 = 841$ smallest
$29^2 + 1 = 842$
$29^2 + 2 = 843$
⋮
$30^2 - 1 = 899$ largest

8. A number has exactly five factor pairs. If the number is less than 100 and larger than 50, what is the number?

81

9. Find the prime factorization for the following numbers.

a) 90	b) 71	c) 441	d) 246
$2 \times 3 \times 3 \times 5$	prime	$3 \times 3 \times 7 \times 7$	$2 \times 3 \times 41$

10. a) What is the least common multiple of 6 and 15? 30

b) What is the greatest common factor of 6 and 15? 3

11. a) The least common multiple of a number and 12 is 60. What could the number be?

5, 10, 15, 20, 30

b) The greatest common factor of a number and 12 is 4. What could the number be?

4, 8, 16, 20, 28 and so on

12. a) In checking for all the primes less than 180, how far do you have to sift?

$\sqrt{180} \approx 13$

b) List all the primes between 200 and 220.

211, 217

NAME

Review Problems

13. To find all the primes from 1 to *N*, you need only sift through 13.

a) What is the smallest number *N* could be? $13^2 = 169$

b) What is the largest number *N* could be? $17^2 - 1 = 288$

14. In paper pool, a certain table has dimensions 6×15.

a) What pocket will the ball end up in? C

b) How many hits are there? 7

c) How many squares are crossed? 30

15. If a pool table has a bottom edge of 12, give the dimensions of the side edge so that

a) the ball ends up in pocket B. 1, 2, 3, 5, 6, 7, and so on

b) the ball has at least seven hits. 1, 2, 5, 7, 9, 10, 11, 13, and so on

c) the ball crosses at least 30 squares. 9, 18, 36, and so on

16. Draw all the rectangles with an area of 18.

17. Every 4th day, pizzas are served on the school menus. Carrot salad is served every 6th day. If carrot salad and pizzas are served today, when is the next time they will be served together?

12 days from now

18. 150 baseball cards and 100 football cards are to be distributed evenly among a group of scouts. What are the possible numbers of scouts for this distribution to come out evenly?

1, 2, 5, 10, 25, 50

Unit Test Answer Key

1. A
2. B
3. E
4. A
5. D

6. C
7. C
8. C
9. E
10. E

11. D
12. D
13. C
14. A
15. D

16. B
17. B
18. E
19. D
20. B

21. C
22. C
23. B
24. D
25. B